The Passing of FutureGen: How the World's Premier Clean Coal Technology Project Came to be Abandoned by the Department of Energy

Report by the Majority Staff of the Subcommittee on Investigations and Oversight of the
Committee on Science and Technology to
Chairman Bart Gordon and
Subcommittee Chairman Brad Miller.

March 10, 2009

The Passing of FutureGen:
How the World's Premier Clean Coal Technology Project Came to be Abandoned By the Department of Energy

Executive Summary

When President George W. Bush announced the FutureGen initiative in February of 2003, he described it as a 10-year, $1 billion, government/private partnership to build a coal-based, zero-emissions electricity and hydrogen producing power plant. It would provide the American people and the world with advanced technologies that would help meet the world's energy needs, and would improve the global environment for future generations. Spencer Abraham, then-Secretary of the Department of Energy (DOE), went even further. This "bold step" would turn coal from an "environmentally challenging energy resource into an environmentally benign one" and demonstrate the best technologies the world had to offer.

The plant would not use traditional coal technology, but would be an integrated gasification combined cycle/carbon capture and storage (IGCC/CCS) facility built at the commercial scale of 275 megawatts. It would sequester one million metric tons of carbon dioxide per year, produce both electricity and hydrogen as energy sources and demonstrate the integration of commercial and untested technologies. Its results would be shared with all participants, including international parties, industry, the environmental community and the public. International participation was a core component of the project as acceptance of the project's results were deemed necessary by the Administration for building an international consensus on the role of coal and carbon sequestration in addressing global climate change and energy security.

But in December of 2007, after a site in Illinois was selected by FutureGen's private industrial partners, the environmental impact statement required by the National Environmental Policy Act was completed, and the State of Illinois had accepted liability for the sequestration aspect of the project, then-DOE Secretary Samuel Bodman announced that he intended to restructure FutureGen. He would "maximize" the private sector role and prevent further cost escalation. The restructured FutureGen was rolled out at the end of January of 2008, but it was widely viewed as the death of the Bush initiative. Subsequent events have verified that view, as the four applications—two of which have been deemed ineligible—responding to the new competition bear no resemblance to the original FutureGen and have no capability to meet the original goals.

How did such a highly publicized Presidential initiative fail, and what were its consequences? Committee staff review of thousands of documents produced by the Department of Energy over the past several months[1] has resulted in the following conclusions:

---

[1] DOE was extremely reluctant to produce documents to the Committee so that it could determine exactly how decisions were made concerning FutureGen. Despite numerous requests from the Committee since April 2, 2008, and the threat of a subpoena, the Department has still not yet provided a full response. Many of the withheld documents involve communications with the White House and this situation has required repeated meetings to examine those materials. We should add that Undersecretary Albright routinely destroyed his e-mail records, further complicating the ability of the staff to reconstruct the full history on decision-making regarding FutureGen.

1.   Based on how easily the Department of Energy abandoned the FutureGen project, it appears that President Bush, Secretary Bodman and the Office of Management and Budget were never fully committed to the FutureGen project or its goal of developing technology to allow the use of coal without massive emissions of carbon dioxide and other greenhouse gases and pollutants. In retrospect, FutureGen appears to have been nothing more than a public relations ploy for Bush Administration officials to make it appear to the public and the world that the United States was doing something to address global warming despite its refusal to ratify the Kyoto Protocol.[2] When worldwide construction costs went up across the board, neither the White House nor DOE was willing to make the additional financial commitment necessary to keep the project going. Secretary Bodman, in particular, strongly disliked FutureGen, and neither President Bush nor any of his White House staff did anything to stop Bodman from killing the original project or restructuring it in a way that was guaranteed to fail. As an assistant to Undersecretary Bud Albright put it during a discussion of restructuring FutureGen:

> "[E]veryone is conveniently forgetting that we're here b/c [because] S-1 [Secretary Bodman] wants to kill FG as its [sic] currently contemplated with or without a Plan B."[3]

2.   Bodman's primary stated reason for killing the original FutureGen plan was that the cost had doubled to $1.8 billion. That was false, and an inexcusable error for the head of a federal agency. Bodman and his staff obtained that number by comparing the cost estimate of $952 million in constant FY 2004 dollars with the "as spent" dollars – which is always higher because it includes normal inflation and other cost increases – that all federal agencies use when estimating the actual cost of multi-year projects such as FutureGen. The Office of Fossil Energy attempted numerous times to explain to DOE's policy staff the difference between these two numbers, but as Under Secretary Bud Albright's chief of staff cavalierly explained while preparing talking points for Bodman, "this is not a legal document, it is a communications document. As for whether the escalation costs after 2004 were expected or not, why does that even matter?"[4]

It is difficult to believe that anyone working at the top levels of DOE or the White House, both of which deal with many multi-year clean-up, research and defense projects – particularly someone with Bodman's business background – did not know the difference between "constant" and "as spent" dollars or even ask how the $1.8 billion figure was obtained. But there is no evidence that anyone asked that basic question.

3.   Secretary Bodman should have known that his claims that the restructured FutureGen would accomplish all of the goals of the original plan and would speed the use of CCS technology were false   Bodman and his senior deputies—Deputy Secretary Clay Sell and

---

[2] . FutureGen was touted as a key climate change inspired action to the Committee on Science in a hearing on September 20, 2006, "Department of Energy's Plan for Climate Change Technology Programs." The Departmental witness stated that "CCTP's portfolio includes realigned activities as well as new initiatives, such as the President's Advanced Energy and Hydrogen Fuel Initiatives, carbon sequestration, and FutureGen," p. 21.

[3] E-mail from Doug Schwartz to Julie Ruggiero, December 10, 2007.

[4] E-mail entitled "Fw: Updated FuturrGen Talking Points" from Doug Schwartz to Andrew Patterson, Dec. 15, 2007.

Undersecretary Albright—demanded that DOE staff create documents for the White House saying the new plan would cost less taxpayer money and do more to validate new carbon capture and sequestration technologies in a shorter time frame than the original FutureGen. This work was largely overseen by political appointees working under Sell and Albright. These claims were concocted without consulting the industry that was expected to take up the FutureGen mantle and despite the repeated warnings of career DOE staff to the political leadership of the Department that the project would fail to meet the original goals. Career staff produced a summary analysis by December 2007 that was entitled, "What "Plan B" would NOT accomplish" (emphasis in original). The concluding paragraphs are so compelling that they are worth quoting at length:

> Given the above delays [following analysis of how Plan B would slow technology development and deployment], it is reasonable to assume that proceeding with "Plan B" and without FutureGen, the availability of affordable coal fueled CCS plants would be delayed at least 10 years and will not allow widespread deployment of CCS until near 2040. Affordable CCS technologies will not be available in time to meet the expected turnover of the existing fleet of coal power plants in the US, nor for incorporation into the development of the world's massive coal resources in countries such as China and India.

> Based on the DOE Climate Change Task Force analysis, which was the basis for the FY09 DOE budget request, a delay of ten years in the deployment of fossil technology with CCS would result in a cumulative loss of emission reductions of about 22 billion tons CO_2 through 2100 in the U.S. To put this into perspective, current U.S. total annual CO_2 emissions are 6 billion tons; U.S. annual CO_2 emissions from coal are 2 billion tons. The DOE Task Force further estimated that CCS benefits from the proposed initiative for the rest of the world were about 6 times the U.S. benefits, or on the order of 150 billion tons CO_2 through 2100 worldwide that would not be avoided if "Plan B" were chosen.[5]

4.   The anemic response by industry to the competition to participate in the new FutureGen proved in a real world demonstration how wrong Bodman and his deputies were. There were four responses of which two were ineligible and two were incomplete. None proposed to construct the IGCC/CCS, coal-based, zero-emission electricity and hydrogen producing power plant that had been promised by Secretary Bodman in January of 2008. The industry response to a Request for Information and the draft FOA had reduced the restructured program to a competition for technology that would attempt to sequester a smaller amount of carbon dioxide, either as part of a newly constructed plant or as a "bolt on" to an existing plant.

---

[5] Analysis from a one page document drawn from e-mails circulating in the Department dated December 11, 2007. These findings were also quoted by Victor Der in an e-mail that went to James Slutz and others in this same time frame, but similar points had been raised by DOE staff throughout the discussion of whether there was a viable option to the President's FutureGen program.

But by the time the career staff were proven right, Bodman and President Bush were at the end of their tenure, the scheduled project selection date had passed, and the United States had lost a year, at minimum, in developing and deploying carbon capture and sequestration technologies.

5.   The Bush Administration's abrupt cancellation of the original FutureGen without bothering to consult or even warn the four countries (India, Australia, South Korea, and China) which had signed on as project partners severely damaged the United States' reputation as an international science partner. The South Korean Minister for Commerce, Industry and Energy wrote on February 4, 2008 (three days after receiving a cancellation notice from Secretary Bodman):

> "I am really surprised that I had no prior explanation of that restructuring intention from DOE... If you have recognized all Korea's endeavor regarding the project, it is not the appropriate way to deliver US DOE's intention to restructure FutureGen by sending me an e-mail."[6]

Foreign partners weren't the only ones surprised by DOE's change of direction. Cancellation of the project, and the abandonment of the growing coalition of countries supporting the project, also allowed the technology lead in this important endeavor to move to other countries. Carbon capture and sequestration projects are now going forward in Australia, China (former partners) and Europe. Other countries no longer look to the United States for leadership in this area, and, as senior DOE officials acknowledged to one another, the restructured program had no international component built into it.[7]

6.   Creating "clean coal" is an extremely complex task involving not only the development of reliable and economical technology to capture carbon dioxide and other pollutants, and integrating it into electricity-producing coal plants, but also the acceptance of higher electricity prices and unknown liability for carbon dioxide sequestration sites by the public and their elected officials worldwide. Without a carbon regulation structure in place, it is almost impossible to expect power generators and utilities to take on this "public benefit" task without expecting a return on investment, something that the Bush Administration refused to acknowledge, much less address. This guaranteed that Secretary Bodman's efforts during the summer and autumn of 2007 to convince industry to sign up for more risk in the original FutureGen project would be a non-starter. FutureGen was a high-risk effort to develop and demonstrate innovative technologies for carbon capture and sequestration. Without a regulatory environment requiring firms to use such technologies, there was little reason – beyond calculations of public relations – for private companies to commit any more than they already had on FutureGen.

When the Department of Energy's top managers were attempting to restructure FutureGen, a senior career official from the Office of Fossil Energy described the new project as

---

[6] . E-mail entitled "Re: DOE Announces Restructured FutureGen" from Kijune Kim to James Slutz, Feb. 4, 2008.
[7] E-mail entitled "RE: Int'l aspect of the new futuregen construct" from James Slutz to Karen Harbert, Dec. 12, 2007.

a Frankenstein.[8] The analogy to the creation of a monster which could not be controlled by its creator was not quite accurate. But the idea that "Plan B" was a cobbled together mess of left-over parts was not far off the mark. However, what DOE really created was more of a Humpty Dumpty.   Just like Humpty Dumpty, when FutureGen fell off the wall in its "restructured" form, it broke apart, and all of DOE's press releases and PowerPoint presentations couldn't put it back together again.[9]

---

[8]. E-mail from Victor Der to Jay Hoffman and Jarad Daniels, January 2, 2008 forwarding the Plan B Program Plan. Der writes in full: "Here's the Frankenstein. I'll be calling NETL [National Energy Technology Laboratory] to see where they are in the electrodes development to make it walk."

[9] Humpty Dumpty's ability to create new meanings for words in Lewis Carroll's *Through the Looking Glass* also bears some relationship to Secretary Bodman's attempt to create something new while still calling it "FutureGen" so that, technically, he could say the President's initiative was alive. "When I use a word," Humpty Dumpty said in a rather a scornful tone, "it means just what I choose it to mean – neither more nor less." "The question is," said Alice, "whether you can make words mean different things." "The question is," replied Humpty Dumpty, "which is to be master – that's all."

The Origins of FutureGen

In his State of the Union address in January of 2003, President George W. Bush unveiled his "Hydrogen Fuels Initiative," otherwise known as a hydrogen-powered, noxious emissions-free car called the "Freedom Car." He committed $1.7 billion over the next 10 years for research on car technology and fuel distribution. But where would the hydrogen fuel come from? In the volume required by the transportation sector, it could only come from coal or natural gas.[10] And thus was born FutureGen.

A month later, on February 27, 2003, the President announced with great fanfare the Integrated Sequestration and Hydrogen Research Initiative, a 10-year, $1 billion, government/private partnership to build a coal-based, zero-emissions electricity and hydrogen producing power plant. "This demonstration project and the Carbon Sequestration Leadership Forum will build on these initiatives to provide the American people and the world with advanced technologies to meet the world's energy needs, while improving our global environment for future generations," he promised.[11] "It will be the cleanest fossil fuel-fired power plant in the world," a contemporaneous Department of Energy (DOE) publication claimed and was a "direct response to the President's Climate Change and Hydrogen Fuels Initiatives."[12] According to then-DOE Secretary Spencer Abraham, the project would "help turn coal from an environmentally challenging energy resource into an environmentally benign one."[13] It would be "one of the boldest steps our nation has taken toward a pollution-free energy future. . . .The prototype power plant will serve as the test bed for demonstrating the best technologies the world has to offer," Abraham promised.[14]

The announcement was made jointly by the Department of Energy (DOE) and the Department of State to emphasize the core objective of international cooperation. At the same time, the two agencies announced the creation of the Carbon Sequestration Leadership Forum (CSLF), an international panel which would focus on carbon capture and sequestration.[15] All these initiatives were in large part a response to President Bush's desire to show that the United States was engaged in efforts to reduce global warming even though it had refused to ratify the Kyoto Protocol because of the generous greenhouse gas emission limits for developing countries. They were hailed by the business press as a "viable alternative to Kyoto."[16]

The 275-megawatt, prototype zero emissions plant subsequently known as "FutureGen" would be a "living laboratory" to test new clean power, carbon capture and coal-to-hydrogen technologies. The DOE release went on to say that President Bush had already emphasized the importance of technology in stabilizing greenhouse gas concentrations in the atmosphere with two major previous policy announcements: the National Climate Change Technology Initiative

---

[10] "A Car for the Distant Future," *The Washington Post*, March 9, 2003, B2.

[11] "Bush Administration Announces $1 Billion Coal Plant Project," *Platts Coal Outlook*, March 3, 2003, p. 1.

[12] "A Vision for Tomorrow's Clean Energy," U.S. Department of Energy, Office of Fossil Energy, February 2003, p. 1.

[13] "U.S. Seeking Cleaner Model of Coal Plant," *New York Times*, Feb. 28, 2003, A22.

[14] "DOE Aims for 'pollution-free' Plant," *Inside Energy/Federal Lands*, March 3, 2003, p. 1.

[15] DOE, "Concept Paper on International Participation in FutureGen," June 2008.

[16] "The Post-Kyoto Initiatives," http://www.allbusiness.com/mining/oil-gas-extraction-crude-petroleum-natural/718535-1.html, Dec. 22, 2003.

on June 11, 2001, and the Global Climate Change Initiative on February 13, 2002. "Carbon capture and sequestration technologies likely will be essential to meeting the President's goals. Without them, it will be virtually impossible to limit global carbon emissions," DOE stated.

Moreover, the President's Hydrogen Fuels Initiative envisioned "the ultimate transformation of the nation's transportation fleet from a reliance on petroleum to the use of clean-burning hydrogen," DOE said. Although most hydrogen in the United States and about half of the world's hydrogen supply were currently produced from natural gas, "The new technologies to be integrated into the prototype plant will expand the options for producing hydrogen from coal, providing a more diversified and secure source of feedstocks for the President's initiative" (emphasis added).[17]

Virtually every aspect of the prototype plant would employ cutting-edge technology. It would not use "traditional coal technology," but be based on a coal gasification system to produce hydrogen and carbon dioxide. The hydrogen would be used for electric power generation or as a feedstock for refineries. "In the future, as hydrogen-power automobiles and trucks are developed as part of President Bush's Hydrogen Fuels Initiative, the plant could be a source of transportation-grade hydrogen fuel." New technologies would be used to capture the carbon dioxide, and it would be sequestered in a geologic formation that would be intensively monitored to verify the permanence of the storage.[18]

The goals of the project were extremely ambitious. DOE and its partners were to:

1.   Design, construct and operate a 275-megawatt prototype plant that produced electricity and hydrogen with near-zero emissions. The size of the plant was driven by a need to provide commercially relevant data and produce 1 million tons of carbon dioxide (CO_2) necessary to validate the "integrated operation of the gasification plant and the receiving geologic formation."

2.   Sequester at least 90 percent of the CO_2 emissions, prove the effectiveness, safety and permanence of the sequestration and establish standardized technologies and protocols for CO_2 measuring, monitoring and verification.

3.   Validate the engineering, economic and environmental viability of "advanced coal-based, near-zero emission technologies" that by 2020 would produce electricity with less than a 10 percent increase in cost; and produce hydrogen at $4 per million Btus or less than the wholesale price of gasoline.[19]

The industry and the environmental community expressed skepticism from the outset. Coal gasification to produce electricity is "still an edgy technology," one expert said, and

---

[17] All discussion of "DOE Release" is from "A Vision for Tomorrow's Clean Energy," U.S. Department of Energy, *supra.* p 1. President Bush reiterated his support for FutureGen in fact sheets and statements related to his administration's environmental and energy accomplishments in October 2003, April 2004, March and June 2005, February and March of 2006, and January, April, May and September of 2007. New foreign partners were welcomed at the White House. "Statements about FutureGEN," undated DOE document.
[18] *Ibid.*
[19] "A Vision for Tomorrow's Clean Energy," *supra*, p. 2.

extracting hydrogen from coal wasted 30 percent of the fuel's latent energy. The budget and schedule were viewed as tight "even for a conventional coal-fired power plant." One environmentalist said until the administration supported a "binding program" to limit carbon emissions, the private sector would not commit "real money" to solving the problem.[20] But if the project reduced the cost of carbon dioxide sequestration from $100 to $300 per ton to $10 or less, it would save the U.S. "trillions of dollars" to meet the inevitable carbon regulations.[21]

By the end of 2003, DOE's Office of Fossil Energy (FE), which had the lead on the project, had prepared the mission need statement required for the acquisition of a capital asset. It focused on the necessity to integrate the operation of a coal-based hydrogen/power facility with carbon dioxide sequestration, something that the existing clean coal research program – which addressed the development of components and subsystems – did not do. To sufficiently consider the feasibility of the zero-emissions concept, DOE had to address the integration gap "to prove technical operational viability to the conservative coal and utility industry."[22] The expectation was that FutureGen would be sufficiently successful that when the aging fleet of coal plants was retired in the 2020-2040 time frame, there would be a viable zero emissions coal option.[23]

In the need statement, FE evaluated and rejected six alternative approaches to achieve President Bush's goal. In particular, it rejected the option of a large-scale demonstration of commercial technology by the power industry. "This alternative would require the immediate integration of a number of complex commercial-scale power plant component technologies, and operation and integration will be technically challenging and risky from an industry perspective." Moreover, the sequestration had not yet been demonstrated. Such an approach would not be cost-effective and without legislated carbon constraints, "the industry has no incentive to invest its limited capital in this demonstration and pursue this high-risk course of action."[24]

The acquisition strategy for a research and development project was conditionally approved by DOE's deputy and undersecretaries in November of 2003 and fully approved in April of 2004. Congress provided $9 million to initiate FutureGen, but also asked for a report on funding and cost sharing.[25] The goals and the administration's plans for achieving them were more fully outlined in the program plan submitted to Congress in March of 2004 as required in the Department of Interior and Related Agencies Appropriation Act of 2004 (P.L. 108-108). The cost share would be 74 percent government and 26 percent private—well above the 20 percent commitment from the private sector normally required for research and development projects.[26]

---

[20] *Ibid.*

[21] "A Pollution-free Coal Plant?" *Power Magazine*, May 2003

[22] "Mission Need Statement: FutureGen Sequestration and Hydrogen Research Plant," DOE Office of Fossil Energy, Nov. 6, 2003, pp. 1-2.

[23] *Ibid.*, p. 4.

[24] *Ibid.*, pp. 12-13.

[25] E-mail entitled "RE: FW: FutureGen Acq Strategy" from Keith Miles to Patrick Ferraro, Feb. 27, 2007.

[26] DOE, Office of Fossil Energy, "FutureGen: Integrated Hydrogen, Electric Power Production and Carbon Sequestration Research Initiative: Energy Independence through Carbon Sequestration and Hydrogen from Coal," March 2004; Conf. Rep. 108-330, 149 *Cong.Rec.* 9898, 9936, Oct. 28, 2003.

In the plan, DOE told Congress that FutureGen "directly" addresses one of the four strategic goals in its 2003 Strategic Plan: to protect national and economic security by "promoting a diverse supply and delivery of reliable, affordable, and environmentally sound energy." Through use of efficient generation technologies and carbon sequestration, FutureGen would eliminate environmental barriers and enable the continued use of domestic coal. It would also produce hydrogen for transportation to support President Bush's hydrogen fuel initiative and provide a "unique real-world opportunity to prove the feasibility of large-scale carbon sequestration, a key potential strategy to reduce the risks of climate change." Absent this "zero-emission option . . ., coal's contribution to the Nation's energy mix could be severely curtailed, thus limiting the fuel diversity of our electricity supply portfolio, and increasing our dependence on more expensive and less secure sources of energy."[27]

Defined as a "public benefits-driven" investment in "high-risk, high-return technology that private companies alone cannot undertake" FutureGen's integration of concepts and components would be the

> key to proving technical and operational viability to the generally conservative, risk-adverse coal and utility industries. Integration issues such as the dynamics between upstream and downstream subsystems . . . can only be addressed by a large-scale integrated facility operation. Unless the production of hydrogen and electricity from coal integrated with sequestering carbon dioxide can be shown to be feasible and cost competitive, the coal industry will not make the investments necessary to fully realize the potential energy security and economic benefits of this plentiful, domestic energy resource (emphasis added).

FutureGen would combine high-risk research activities, advanced generation coal gasification technology integrated with combined cycle electricity generation, hydrogen production, and carbon capture and sequestration. It would take at least 10 years to accomplish its goals, and the results would be shared with participants, industry, the environmental community, international partners and the public. "Broad engagement of stakeholders early on in FutureGen is critical to achieving an understanding and acceptance of sequestration and zero-emission coal utilization," DOE stated.[28]

While its goals and schedule were recognized as aggressive and high-risk, they were judged achievable and would prove "the basis for a potentially huge long-term public benefit." And DOE determined that it was not possible "to reach FutureGen's stretch goals using off-the-shelf commercial technology." Critical components needed to be designed, and their efficiencies, environmental performance reliability and economics needed to be advanced and tested. More importantly, "[a] key piece of FutureGen is proving the viability of sequestration and its integration with a power facility."[29] Full-scale operation with continuous power generation was projected by FY 2012.[30]

---

[27] DOE, "FutureGen: Integrated Hydrogen, Electric Power Product ion and Carbon Sequestration Research Initiative," *supra*, p. 2.
[28] *Ibid.*, p. 3
[29] *Ibid.*, p. 6.
[30] *Ibid.*, p. 13.

Furthermore, according to White House officials, the hydrogen transportation initiative and FutureGen were investments that would achieve "both goals of addressing climate change and protecting our economy."[31]

In 2005, after *The New York Times* alleged that industry would not spend money to reduce emissions under a voluntary system that gave a competitive advantage to those companies that did nothing, Samuel Bodman, the new DOE Secretary, reiterated the Department's support for FutureGen.[32] President Bush also featured it prominently in a 2005 "fact sheet" concerning how he was addressing climate change. In December of 2005, Bodman announced an agreement with an industry consortium called the FutureGen Industrial Alliance, to build FutureGen, "a prototype of the fossil-fueled power plant of the future." He described it as a direct response to President Bush's directive to develop a hydrogen economy by "drawing on the best scientific research to address the issue of global climate change." Bodman lavishly praised the Alliance members, who would contribute $250 million to the project, as among "the world's most responsible and forward thinking coal and energy companies." At the heart of the project – described as a "stepping-stone toward future coal-fired power plants" – would be coal-gasification technologies that could eliminate air pollutants and mercury. Carbon sequestration would be a key feature with the goal of capturing 90 percent of the plant's carbon dioxide emissions. The "ultimate goal for the FutureGen plant is to show how new technology can eliminate environmental concerns over the future use of coal and allow the nation to tape the full potential of its coal reserves," Bodman said.[33]

By January of 2006, the project now known as FutureGen was no longer being promoted as a source of transportation-grade fuels, perhaps because the Administration had realized that commercially viable hydrogen-powered cars were some decades away.[34] FutureGen was now to integrate advanced coal gasification technology, hydrogen from coal, power generation, and carbon dioxide (CO_2) capture and geologic storage. "The success of FutureGen will assure that coal, a low-cost, abundant, and geographically diverse energy resource, continues to globally supply exceptionally clean energy."[35]

---

[31] Statement of James Connaughton at Oct. 22, 2004, "Ask the White House," http://georgewbush-whitehouse.archives.gov/ask/20041022.html

[32] "Climate Change and the President," letter from Secretary Bodman, *The New York Times,* May 26, 2005, responding to "Dirty Secret: Coal Plants Could Be Much Cleaner," May 22, 2005. That article referred to the recommendation of the National Commission on Energy Policy, an independent, bipartisan advisory body that the government spend an additional $4 billion on IGCC technology over 10 years to speed up the industry's acceptance of the technology.

[33] "FutureGen Project Launched: Government, Industry Agree to Build Zero-Emissions Power Plant of the Future," DOE press release, Dec. 6, 2005. There were ultimately 13 industrial partners of which four were foreign-based: American Electric Power Service Corp., Anglo American Services Ltd., BHP Billiton Energy Coal, Inc., China Huaneng Group, Consol Energy, Inc., E.ON U.S. LLC, Foundation Coal Corp., Luminant, Peabody Energy Corp., PPL Energy Services Group, Rio Tinto Energy America Services, Southern Company Services, Inc., and Xstrata Coal Pty Ltd.

[34] "When Presidents Talk Fuel, the Nation Listens, Sort Of," *Detroit Free Press*, Feb. 13, 2006, B2.

[35] DOE, "FutureGen – A Sequestration and Hydrogen Research Initiative," Project Update: January 2006."

The project appeared to be going well in this time frame – at least publicly. A preliminary agreement with the Alliance was signed on December 2, 2005.[36] President Bush referred to it in his 2006 State of the Union address as part of his Advanced Energy Initiative.[37] Participation by foreign governments was expected.[38] Its cost in FY2005 constant dollars was $952 million.[39] According to DOE's assistant secretary for fossil energy, "the FutureGen project is being pursued aggressively and is on schedule."[40] It was a "high priority," James Connaughton, chairman of the White House Council on Environmental Quality and the President's senior environmental and natural resources adviser, stated in late 2006.[41] By April of 2007, a first phase cooperative agreement had been signed which would include work on siting, scoping, conceptual design and National Environmental Policy Act (NEPA) compliance. The Alliance had selected four sites as finalists, and the winning site was expected to be announced in mid- to late 2007.[42]

The significance of the FutureGen project on the international stage could not be underestimated. After his refusal to submit the Kyoto Protocol to the Senate for ratification, President Bush and his advisers touted the highly visible project as a way to attack the problem of global warming in the voluntary, cooperative international manner that was a hallmark of the Bush approach to environmental problems. CEQ Chairman Connaughton, who had the task of defending the Bush administration, did so by promoting international partnerships for sustainable growth, of which FutureGen was one.[43] It was particularly important in U.S. relationships with India and China, both of which signed on as partners in the FutureGen project even before the cooperative agreement with the Alliance was completed. A "U.S.-India Energy Dialogue" was established by Secretary Bodman and Montek Singh Ahluwalia, deputy chairman of India's Planning Commission, in 2005. By May of 2006, India had become the first foreign country to sign on as a FutureGen partner. According to Senate testimony in 2007 by David Pumphrey, then DOE deputy assistant secretary for international energy cooperation, "successfully demonstrating and adopting this technology will allow India to reduce the intensity of future greenhouse gas emissions from the burning of their abundant coal resources."[44]

In September of 2006, President Bush and President Hu Jintao of China agreed to create a "Strategic Economic Dialogue" (SED) between the two countries which would be convened

---

[36] DOE, "FutureGen Status," PowerPoint presentation for 7th annual SECA Workshop and Peer Review, Sept. 12-14, 2006.

[37] In a press release providing a more detailed description of the initiative, the Administration noted that the 2007 budget included $54 million for FutureGen as part of the clean coal technology program. The White House, "State of the Union: The Advanced Energy Initiative," Jan. 31, 2006, p. 1.

[38] British, Australian and Chinese companies were already Alliance members. http://www.futuregenalliance.org/alliance/members.stm Four countries (India, Korea, Japan and China) also joined.

[39] Constant dollars are not an accurate reflection of the actual cost of a 10-year, lifetime project over the life of the project because they do not include cost increases that result from inflation and changes in construction, materials and other costs during the out-years. In its 2004 report to Congress, DOE did not point out that it was using constant year dollars when projecting the total cost of the project. DOE, "FutureGen: Integrated Hydrogen, Electric Power Production and Carbon Sequestration Research Initiative, *supra*, p. 9, Figure 3.

[40] "Clean Energy Project," letter from Jeffrey Jarrett, *The New York Times*, June 5, 2006.

[41] "Budgets Falling in Race to Fight Global Warming," *The New York Times*, Oct. 30, 2006, A1.

[42] *Ibid.*, p. 2.

[43] "Bush Aide Says Myths about US' Green Policy Remain," *The Economic Times*, Aug. 30, 2006.

[44] Statement of David Pumphrey before the U.S. Senate Committee on Energy and Natural Resources, July 18, 2006, p. 4.

semi-annually. Treasury Secretary Henry Paulson would lead the U.S. side of the dialogue, and the Energy Department would dialogue with China's National Development and Reform Commission on energy policy.[45] In December of 2006, China – the second largest producer of CO_2 after the U.S. – became the third foreign country (South Korea was the second) to join the FutureGen Government Steering Committee. China Huaneng Group, the country's largest coal-fueled power generator, had already joined the Alliance. According to Pumphrey, the U.S. "assigned a high priority to maintaining long term technical cooperation with China on fossil energy issues," including FutureGen. The FutureGen concept could demonstrate technologies that would reduce carbon emissions worldwide.[46]

The Cost Issue

By early 2007, however, DOE management internally was raising questions about the cost of FutureGen. Even before the Full Scope Cooperative Agreement was signed, DOE headquarters was expressing its discontent to the Alliance. FutureGen's as-spent cost projection, which included inflation and the increasing cost of construction and materials, was $1.8 billion and global construction costs were rising. In light of those anticipated cost increases, DOE was balking at paying 74 percent of any additional costs even though an increase in as-spent costs would normally be expected. Michael Mudd, the Alliance's chief executive officer, expressed his concern about DOE's delay in signing the cooperative agreement, saying it would cause schedule and engineering delays and a loss of credibility. "We do not understand why issues, such as the cost-share fraction, continue to be revisited. This specific issue was settled nearly two years ago during discussions between the White House, OMB, DOE, and the Alliance." The Alliance would like to report "positive progress" on all fronts to Congress "rather than concerns that the Administration is having second thoughts about supporting the FutureGen project."[47]

In a discussion over a draft press release announcing the agreement, Victor Der, then director of DOE's Office of Clean Coal Systems,[48] complained to George Rudins, former deputy assistant secretary for coal and power systems, that the release emphasized a cost increase, not the fact that "notwithstanding rising inflation in the heavy construction sector, both the Alliance and DOE believe that FutureGen is vitally important to coal and climate change, and have committed to continuing as cost shared partners in this initiative."[49] FE also reminded the Department that it was a "key Presidential Initiative and a major Government/Industry Partnership" for producing electricity and hydrogen from coal while eliminating emissions and sequestering carbon dioxide at a low cost.[50] The final press release did, however, refer

---

[45] "Fact Sheet Creation of the U.S.-China Strategic Economic Dialogue," Treasury Department press release, Sept. 20, 2006.

[46] "US-China Relationship: Economics and Security in Perspective," Statement by David L. Pumphrey before the US-China Economic and Security Review Commission, Feb. 1, 2007, p. 7.

[47] E-mail entitled "FutureGen delays," from Michael Mudd to George Rudins (cc: Carl Bauer, Keith Miles, Thomas Russial, Thomas Sarkus) March 20, 2007.

[48] Dr. Der has held various positions at DOE related to fossil energy and clean coal. He is currently acting assistant secretary for fossil energy.

[49] E-mail entitled "Fw: FutureGen release: FE first draft" from Victor Der to George Rudins, March 25, 2007.

[50] E-mail entitled "RE: FG @ Revised Congressional" from Thomas Shope to Dirk Bartlett, William Purvis and Raj Luhar, March 23, 2007.

specifically to the cost increases, but said a review of "progress and expenses" would not be concluded until the end of the first phase of the project in June of 2008.[51]

The Alliance was so upset by DOE's concerns as expressed in a call from Deputy Secretary Clay Sell on the day the press release was issued that Mudd said it was "putting the project on hold until we have the chance to meet with Clay and Secretary Bodman to address issues and concerns raised by Clay during his call."[52] When asked later in a press call why DOE signed the agreement if it already had these concerns, Sell said it was the signing of the agreement that brought the financial issues to his and Secretary Bodman's attention.[53]

Sell and Bodman did not waste any time bringing their hesitation to the White House. In April, Sell briefed staff of the National Economic Council, OMB, the National Security Council and the Office of the Vice President on their cost concerns, and it was agreed that the costs had to be capped.[54] Thomas Shope, DOE's principal deputy assistant secretary for fossil energy, communicated to the Alliance that "the project will not move forward as currently structured." Within days, DOE's lawyers were asked to determine if the agreement made clear that DOE could "just decide not to fund it if it got too expensive" or how to cap its contribution.[55]

At a May 11, 2007, meeting with NEC and OMB staff, Shope recorded the following:

DECISIONS: The significance of the project in the Administration's global climate change strategy was recognized. However, additional cost containment measures must be part of the project going forward and must be negotiated before the commencement of BP-2. The principal cost containment measure employed will be a cap on DOE's expenditures.[56]

The $1.8 billion as-spent figure had been obtained by adding a straight-line 5.2 percent annual escalation factor during the construction of the contract to the FY 2004 estimate of $950 million, a normal process for all large projects built over a number of years. The Alliance then subtracted $301 million in estimated income from the sale of electricity to come up with a net cost of $1.46 billion. FE staff accepted that as a reasonable escalation, but construction costs in early 2007 were growing at a much higher rate because of worldwide demand for construction services and materials.[57]

---

[51] "DOE Signs FutureGen Cooperative Agreement," States News Services, April 10, 2007; "Rising Costs of FutureGen Plant Heighten Concerns among Legislators," *Platts Coal Outlook*, April 16, 2007.

[52] E-mail entitled "Re: FutureGen Agreement" from Michael Mudd to John Grasser, April 11, 2007.

[53] Transcript of Department of Energy conference call, Jan. 30, 2008. The speakers were Sell and Secretary Bodman.

[54] E-mail entitled "Re: Futuregen...problems" from Jeff Kupfer to Clay Sell, Sept. 9, 2007.

[55] E-mail from Thomas Shope to Clay Sell and Dennis Spurgeon, April 19, 2007; e-mail from Mary Egger to Gene Cadieux, April 16, 2007.

[56] "Meeting Notes 'To Discuss The Revised Cost Estimates For The Futuregen Project,'" attached to e-mail entitled "FutureGen Meeting Followup" from Thomas Shope to Jeffrey Kupfer, Dennis Spurgeon, Karen Harbert, Eric Nicoll and David Hill, May 11, 2007.

[57] E-mail entitled "Table of RTC Escalated Outlays," from Thomas Sarkus to Victor Der and Jeffrey Hoffman, April 2, 2007.

In an April presentation on the project's status to DOE, Mudd and his team pointedly noted that they "trusted" that DOE still shared the vision the administration had put forward "and planned to provide the political, technical and financial support required." He reminded DOE that the FutureGen Alliance was formed in "direct response" to President Bush's initiative, and that the industry was contributing nearly $400 million with "no expectation of financial return," but believed that FutureGen was central to reducing the cost of addressing climate change by "trillions of dollars." FutureGen was unique as no other fully integrated power plant combined gasification and carbon capture and sequestration in a deep geologic formation. It provided "a clear mechanism to assess the cost, performance, and public acceptance of integrated near-zero emissions power plant, which is an essential precursor to commercial deployment." Mudd also pointed to the global significance of such a project as a catalyst for new projects in other countries and its ability to position the U.S. as a leader on climate change solutions.[58]

Mudd reminded DOE that the Alliance members "came to the table" with certain understandings: the government would pay 74 percent of the cost; it would maintain its support of FutureGen; and that the $950 million cost was in FY 2004 dollars and subject to adjustment for inflation which would be shared. For their contribution, Alliance members would get no financial return or intellectual property rights. At that time, every milestone had been met. Construction would begin in 2009, but Mudd pointed out that heavy construction costs were up by 30 percent and well drilling costs by 250 percent.[59] Work continued through the summer on the design and the environmental impact statement, and DOE continued to solicit foreign partners.[60]

These exchanges marked the beginning of a dual track on FutureGen. The administration continued to unequivocally support FutureGen in public. For example, at the end of the April 2007 U.S. – EU summit on energy security, efficiency and climate change, the White House issued a joint statement pledging its support for FutureGen without reservation. "The United States, in partnership with its government steering group member countries and the private sector, will build FutureGen, the United States' first near-zero emissions fossil fuel plant, by 2012," the statement read. The first priority was deploying "near zero emissions coal technologies" which were critical in tackling global CO2 emissions because of coal's importance in meeting energy needs.[61] FE pushed the general counsel's office to "move out on the EIS [Environmental Impact Statement]" so that final site selection could be completed by the end of 2007 because the states had purchase options on sites that expired at the end of the year.[62]

But inside the DOE leadership, it was a different story. In addition to meeting with White House staff, Deputy Secretary Sell was beginning to discuss the "path forward" with senior DOE officials, specifically on how to deal with the project's cost escalation. At the same time, the agency was preparing its FY 2009 budget. Funds for FutureGen – which did not have a

---

[58] FutureGen Alliance, "FutureGen: Project Status," April 18, 2007, pp. 3-5 and 15.
[59] *Ibid.* pp. 8, 10, and 14.
[60] See, *e.g.,* e-mail entitled "FW: Revised TOC" from Joseph Giove to Carol Loman attaching IEA Ministerial 2007 Briefing Book Tasks, April 17, 2007.
[61] "2007 U.S.-EU Summit Statement: Energy Security, Efficiency, and Climate Change," The White House Press Office, April 30, 2007, pp. 1-2.
[62] E-mail entitled "Fw: FutureGen Meeting Followup" from Thomas Shope to David Hill, May 13, 2007.

specific line item in the budget – had to compete annually with other coal research projects such as the Clean Coal Power Initiative (CCPI) and regional carbon sequestration partnerships.

The Alliance did not want to negotiate a new cost agreement until it had completed more reliable cost estimates at the end of the first phase of the project in June 2008 – as anticipated in the cooperative agreement – when it would have a more definitive design.[63] It responded to the pressure from DOE by appealing directly to President Bush in a letter on June 18, 2007. Describing FutureGen as a "premiere global project" with international partners, Mudd wrote that the Alliance members

> have dedicated to FutureGen staff with global expertise in major design and construction projects, and the venture is operated with the clear objectives and management discipline of any major commercials endeavor. Costs are up for every major energy infrastructure project, but the FutureGen Alliance is watching costs closely as we share in the cost increases.

Mudd reminded the President that "To date, your Administration has supported this important global effort" and referred to Bush's May 31, 2007, call for "expanding global cooperation on research and development to bring to market technology based solutions to climate change concerns." Continued government support of FutureGen was critical as staff had to be hired, land agreements made and major plant components with long manufacturing lead times needed to be ordered.[64]

DOE management was not deterred. By July of 2007, Shope had sent a memo to Secretary Bodman asking for the Secretary's approval of an immediate renegotiation of the final cost structure instead of waiting until June 2008.[65] The Alliance's initial response was that the cost increases were not the fault of anything the Alliance had done or failed to do, and reiterated the commitments the members had made through a non-profit consortium. According to the Alliance, there were already rumors from the foreign Alliance members that the U.S. might not be that committed to FutureGen. Nonetheless, Secretary Bodman approved Shope's proposal on July 27 without addressing the commitment issue.[66]

In an accompanying memo to Sell listing various options, however, Shope said that FutureGen was configured to "precisely" achieve the cost and performance goals for the zero emissions coal program and to gain industry acceptance and commercial deployment of the technology on a domestic and global scale. It also had strong international support as the "premier international, collaborative project" addressing greenhouse gases and climate change. Shope noted that the Alliance had been generally willing to work with the Department on cost overruns attributable to design errors, mismanagement, delays from accidents, etc. But the increases projected did not fall into any of those categories, and Shope was very skeptical that

---

[63] E-mail entitled "Re: FutureGen Mtg," from Victor Der to Raj Luhar, Mr. Giove, George Rudins and Jarad Daniels, May 7, 2007; e-mail entitled "FutureGen Path Forward," from Thomas Shope to Clay Sell, May 7, 2007.
[64] Letter to President Bush from Michael J. Mudd, June 18, 2007. No response to this letter was found in the DOE files provided to the Committee.
[65] "Memorandum for the Secretary" from Thomas D. Shope, July 27, 2007.
[66] "Memorandum for the Deputy Secretary" from Thomas D. Shope, Attachment A to "Memorandum for the Secretary, *supra*, July 27, 2007.

the industry would take on additional risk because there was no direct or immediate return on its investment, and it was risk-averse.[67]

Despite the recognition by DOE of these significant factors pointing to FutureGen as the only way to obtain the cooperation of the coal and power industry , DOE had already determined that it was not "financially sustainable." In an August memorandum to Bud Albright, DOE's undersecretary, Shope also said that the Administration was expressing concerns about the cost, although no documents have been provided to the Committee to verify that statement. However, it was clear that the Secretary's single goal was to limit the federal government's cost.[68]

DOE's plan to renegotiate was discussed with the Alliance staff, who told Sell they would work to resolve the issue before the final site selection at the end of the year, but whose nervousness about DOE's commitment to the project was evident. "The talk on the street that the project is in trouble is affecting [the Alliance's] ability to secure good vendors and competitive bids . . . . The Alliance has been told that some vendors are not interested in chasing after the FutureGen project if it just going to fall apart [sic]."[69] But in late August DOE told the Alliance board that a negotiation team needed to be formed.[70] According to talking points prepared for the meeting, Shope told the Alliance that "an 'open checkbook' approach is unsustainable and sets an unrealistic expectation which needs to be addressed. Simply put, we cannot commit to funding the project regardless of cost." For the Department to continue in the partnership, "the FutureGen financial plan must properly incentivize all parties to control costs and to account for those costs that are not directly controllable."[71]

It was a strange message to the partners that DOE had solicited to join in its risky project – and which everyone at DOE knew did not have much of an incentive to join. DOE was now threatening to pull out of its own project and appeared to be shifting the burden of the project momentum to the Alliance. It was now up to the Alliance to keep FutureGen alive.

In early September, staff at DOE's National Energy Technology Laboratory (NETL)— the managers of the FutureGen project—were told that "SE-1 [Secretary Samuel Bodman] and SE-2 [Deputy Secretary Sell] are directing DOE to 'renegotiate' the FG award, based upon their assessment that it is a 'bad deal.'" NETL was to identify areas for cost reduction.[72] In a preliminary meeting that Sell had with the Alliance, he was told that the Alliance was "potentially interested" in reducing its scope so that option was now on the table.[73] NETL quickly responded. "Anything but minor scope changes now could really screw things up." It

---

[67] *Ibid.*

[68] "Memorandum for the Deputy Secretary" from Thomas D. Shope, Aug. 31, 2007.

[69] Undated memo to Clay Sell. Because of the size of the components for an IGCC plant, the Alliance needed to order parts long before they were actually needed.

[70] "Appendix 2: DOE and FutureGen Alliance Communication Timeline," attached to undated FutureGen strategic plan.

[71] "Talking Points – Meeting with Futuregen Alliance Board of Directors," Aug. 29, 2007.

[72] E-mail entitled "Fwd: Pre-Meeting Tuesday morning on FutureGen negotiations" from Keith Miles to Edward Simpson and Ferraro, Sept. 4, 2007. Miles asked the recipients not to "shoot the messenger" and said he was being "asked to identify a 'soldier' from your shop to participate."

[73] E-mail entitled "FG" from Adam Ingols to Thomas Shope and Andrew Patterson, Sept. 6, 2007.

could mean another site "best and final offer" process, a supplemental draft environment impact statement and perhaps the loss of some foreign contributors. NETL's counsel added, "I would be willing to bet the Alliance wants to reduce the CO2 capture level and eliminate the co-sequestration test. The latter might not be such a big deal. The former could open a pandora's box."[74]

Sell also wrote to CEQ Chair Connaughton, Barry Jackson, who had replaced Karl Rove, and Keith Hennessey, President Bush's chief economic adviser, at the White House, and Stephen McMillin, the Office of Management and Budget's (OMB) deputy director in charge of the federal budget, telling them that FutureGen was heading in a "bad direction." It was experiencing significant cost increases, and DOE might be forced to cancel. Sell said that neither the Secretary nor the OMB had contemplated these expenditures and expressed his belief that FutureGen was becoming a bad deal for the government and politically unsustainable in Congress. Sell said other priorities in coal research were being threatened by FutureGen.[75]

Connaughton, who was the Administration's representative at international meetings on climate change, asked that a "tiger team" be put together on the problem. Pointing out that FutureGen was an important part of the administration's climate change response, Connaughton emphasized, "This project is very important . . . . If there is a rational option, it should be considered."[76] There is no indication that this was done.

Options: Strip Down the Project or Change the Cost Share

As requested, FE had put together the pros and cons for various options. It did not favor any major change in the project scope because that would change the basic goals of the project, reduce international involvement and delay clean coal technology development. Specifically, it found scaling down the plant size from 275 MW to 120 MW, a 60 percent reduction which would reduce the cost by only 33 percent, would not meet industry's needs. It would be inefficient, delay the NEPA process, not meet the goal of sequestering 1 million tons of CO2, and still require a subsequent demonstration in a larger plant. FutureGen's goals would be delayed by five years, and the total cost of the program would increase.[77]

Later in the negotiations, a NETL staffer worried: "It occurred to me that we are beating the process 'integration' drum pretty hard in our justification for FutureGen, but I don't think Jim Slutz and most of DOE top management (or anyone at OMB) have an intuitive feel for what these integration issues are and why dealing with them at large scale is so important. . . . The goal is to drive home the point that these integration issues are real and challenging, and are not

---

[74] E-mail entitled "Re: Fw: FG" from Thomas Russial to Jarad Daniels, Victor Der and Thomas Sarkus, Sept. 7, 2007.

[75] E-mail entitled "Futuregen . . . problems" from Clay Sell to James Connaughton et al., Sept. 7, 2007.

[76] E-mail from James Connaughton to Clay Sell, Barry Jackson, Keith Hennessey and Stephen McMillin, **(date?)**

[77] DOE , "FutureGen Options & Recommendations by DOE FE, October 2007, p. 4.

going to be solved at smaller-scale."[78] Adding CCS to the back end of the system and making certain that all the pieces work in tandem would be a significant challenge.[79]

Another option was to break the project into three separate projects for 1) sequestration, 2) the turbine, and 3) the gasifier. FE described its previous negative experience with such a system and said it would be difficult to find companies to do the individual pieces because there was no economic reason to do so. For example, no one would take over the sequestration piece because there was no revenue resulting from sequestering, burying and monitoring CO2.[80]

Reducing the research and development components of FutureGen, which had been sold as a "living laboratory" to test out new technologies was also rejected. The research was needed to prove that there would be no more than a 10 percent increase in the cost of electricity by adding CCS. Without testing in FutureGen, "advanced R&D components would first need to be proven independently and then proven in an integrated fashion at a commercially relevant scale" which "would significantly delay the availability of the technology for commercial deployment and would increase overall cost to the program."[81]

FE also rejected reducing the carbon capture system efficiency from 90 to 50 percent, reducing fuel flexibility or removing the coal-to-hydrogen component. The only viable option for a successful FutureGen was to renegotiate the cost share and have a firm DOE cap as Secretary Bodman had made it clear that he would not sign on to a $3-$4 billion deal.[82]

In September of 2007, FE made its presentation to DOE Deputy Secretary Sell. Citing once again the benefits of FutureGen in proving advances in power generation in an integrated fashion with a variety of coal types, furthering international cooperation with coal giants China and India and proving the viability of widespread CCS, FE recommended that the project scope remain the same, but that further cost increases be shared 50/50 with the Alliance and title to the plant be given to the Alliance to be used for loan collateral.[83] DOE Under Secretary Albright, the agency's lead on the negotiating team, apparently agreed with FE's analysis.[84]

President Bush seemed unaware of the concerns of DOE management. He continued to tout the original program. On September 4, 2007, he issued a joint statement with then-prime minister John Howard of Australia welcoming Australia to the FutureGen International Partnership, which President Bush described as

---

[78] E-mail entitled "Re: IGCC/CCS Process Integration Made Simple" from Jay Braitsch to Thomas Sarkus, Nov. 7, 2007.

[79] E-mail entitled "Re: IGCC/CCS Process Integration Made Simple" from Thomas Sarkus to Jay Braitsch, Nov. 7, 2007.

[80] *Ibid.,* pp. 5-6.

[81] *Ibid.,* p. 7.

[82] E-mail entitled "Fw: FG" from Victor Der to Thomas Russial, Thomas Sarkus and Keith Miles, Sept. 10, 2007; e-mail entitled "FutureGen" from Bradley Poston to Thomas Brown, Oct. 30, 2007.

[83] DOE, "FutureGen Renegotiation Issues and Recommendations," Sept. 14, 2007. FE's guidance for that meeting was to concentrate on scope reduction costs and benefits, not a change in cost sharing. E-mail entitled "FG Guidance" from Andrew Patterson to Jarad Daniels, Sept. 12, 2007.

[84] E-mail entitled "RE: FG Update & Data Call" from Jarad Daniels to Samuel Biondo, Victor Der and Joseph Giove, Sept. 21, 2007.

a major United States-led international project aimed at building a prototype plant that integrates coal gasification and carbon capture and storage to produce electricity with near-zero emissions. This demonstrates and underscores the commitment of both countries to the development and deployment of clean coal technologies.[85]

Negotiations

The initial negotiation session was held in the first week of October. In that meeting, the Alliance agreed to a 50/50 cost split after the first $1.8 billion, but said it had "cost flow constraints." It was considering financing options to help "smooth" the costs to its members during the construction phase. The Alliance proposed that it receive 100 percent of the program income, and that DOE vest title to the plant in the Alliance at the beginning of the project, instead of the end. DOE found this unacceptable, but said internally that the next round of negotiations would focus on "ways to adjust revenue and cost share with the hope of finding a 'win-win' position."[86]

The financing issue continued, however, to be the critical sticking point. The Alliance wanted to fund the project through a leveraging plan; DOE refused.[87] By the end of October, DOE proposed that the individual Alliance members each give a guarantee "for a significant portion of the financing. If the Alliance defaults or withdraws, the members must pay over the guaranteed amount to the lender to reduce the outstanding debt thereby making it more economically practical for DOE to take over and complete the project."[88]

By early November, DOE told the White House that it had begun work on a "parallel strategy" if no agreement could be reached. It would seek to maintain the goals and objectives of FutureGen by "(a) adopting a different partnership construct that makes more sense for the Federal government, or (b) separating the project's core technologies and accelerating our ongoing R&D efforts in these areas, testing at smaller scale with limited integration, and expediting deployment to the marketplace."[89] It was the beginning of what would be known as Plan B, an idea first mentioned by Bradley Poston in DOE's Office of Contract Management. Poston had asked if costs could be reduced by using an existing power plant to test out the carbon capture and sequestration products still in the research and development stage. Poston concluded that without carbon sequestration, there was no reason to proceed "so either the costs

---

[85] "Statements about FutureGEN," undated document from DOE, p. 1. President Bush also told foreign media in late May of 2007 that he believed FutureGen would be developed as a coal-fired plant with zero emissions. "And when that technology comes to fruition, if you can get yourself some coal, you've got your ability to diversify away from sole-source supplier of energy." Remarks by President Bush in Roundtable Interview with Foreign Media, http://fpc.state.gov/fpc/85918.htm, May 31, 2007. International participation was not that easy to obtain. Prospective contributors weren't sure what they were getting for their $10 million. Because of proprietary concerns, visiting researchers would not be able to fully view certain project areas. If too much information was shared, vendors might not be attracted to the project, DOE worried. Some kind of licensing arrangements might be possible, but they were never worked out. E-mail entitled "Re: FutureGen Renegotiation process update" from Thomas Russial to Jarad Daniels, Sept. 19, 2007.

[86] DOE, "Brief Summary: First Round of Negotiation between FutureGen Alliance & DOE," Oct. 4, 2007, pp. 1-2

[87] Attachment to e-mail entitled "FutureGen Timeline.doc" from Doug Schwartz to Kasdin Miller, Jan. 24, 2008.

[88] E-mail entitled "FutureGen Renegotiation" from Thomas Rusial to David Hill and Mary Egger, Oct. 19, 2007.

[89] E-mail entitled "New final paragraph for futuregen" from Adam Ingols to Sarah Magruder, Nov. 1, 2007

are reduced significantly or we revise our goals and focus on getting most of the technologies developed now so that in the future we can design and build with greater knowledge and confidence in our success and cost control."[90]

Plan B

Top DOE officials soon proposed a new FutureGen structure under which private companies would fund the IGCC plant, and DOE would pay only for the CCS component. In an e-mail exchange with a White House staffer, Albright described FutureGen's current structure as not only fostering cost overruns but actually threatening the "success of the underlying goals of FutureGen."[91] DOE's clean coal research team did not agree. According to FE, the national and global costs of not going forward with the original plan would be enormous. Private industry would not take on this challenge without significant incentives and the passage of carbon reduction legislation that gave a value to carbon. ***"Given the above delays, and assuming a reluctance to pursue high-cost alternative pathways, it is reasonable to assume that without FutureGen, the availability of moderate-cost, coal fueled CCS plants would be delayed by 10-15 years."*** (Emphasis in the original)[92]

The 10-year delay would result in a loss of U.S. emission reductions of about 22 billion tons of CO2; a 15-year delay would result in a loss of 33 billion tons. For the rest of the world, however, the loss of this technological research would be six times the U.S. losses, or about 150 billion tons. Having a stream of commercially available, increasingly cost-effective coal/CCS technology options beginning in 2020 would also reduce electricity and natural gas costs. "Integration of concepts and components in a full scale test facility like FutureGen is the key to proving the technical and operational viability as well as gaining acceptance of the near-zero emission coal concept," staff wrote.[93] In undated notes of an internal discussion, Karen Harbert, the assistant secretary for policy and international affairs, also reminded the group that DOE had gotten a "plus up" in the FE budget by claiming that it would significantly accelerate CCS development by 2030, and that there would be a "big problem" if there was a delay.[94]

These warnings were pushed aside as Albright, Sell and the DOE policy staff moved forward with Plan B. This structure would scrap the cooperative agreement, the Alliance and the international partners for a new competitive procurement under which individual U.S. companies would take on the responsibility of building IGCC plants, and DOE would pay only the additional cost of the CCS component. At the same time, however, DOE continued to negotiate with the Alliance on the cost share and financial component and continued working on the EIS for the four sites which were the finalists.[95] FE raised again the problems with IGCC plants. Only two had been built, and both ran on natural gas, not synthetic gas or hydrogen from coal. "Some of us tekkies worry that hydrogen will pose an even greater challenge than syngas did.

[90] E-mail entitled "FutureGen" from Bradley Poston to Thomas Brown, Oct. 30, 2007.
[91] E-mail entitled "Re: FutureGen Funding" from Bud Albright to Charles Blahous and Clay Sell, Nov. 6, 2007.
[92] "Discussion of Alternative FE Clean Coal Program without FutureGen," p. 2, attached to an e-mail entitled "Re: Alternative FutureGen Plan C" from Thomas Sarkus to Doug Schwartz and Victor Der, Nov. 9, 2007. It appears that DOE briefly considering eliminating FutureGen altogether, but discarded that option.
[93] *Ibid.*, pp. 2-3. DOE, "What 'Plan B' would NOT accomplish, undated.
[94] Undated notes of meeting on Plan B. Participants: Karen Harbert, Victor Der, Scott Klara and Jim Slutz.
[95] E-mail entitled "New final paragraph for futuregen" from Adam Ingols to Sarah Magruder, Nov. 1, 2007.

Add-in a water-gas shift reactor, which no IGCC plant now has. Then tack CCS onto the back end and make certain that all of the pieces work in tandem. You get the drift," a NETL engineer wrote.[96]

Other people started to raise questions, and the scramble was on to justify Plan B. A debate between Doug Schwartz, Albright's chief of staff, and Poston revealed the difficulties of making the new plan viable – even on paper. Poston said a new competition would delay the schedule, and he could see no industry self-interest. "We may give a party no one comes to," he wrote. Schwartz answered that DOE would just have to create more self-interest.

> [T]here may be a new model(s) we come up – in theory – that may alter our prior determination there is no return on investment for partners, whether resulting from changing the IP approach, permitting vendors to participate, an impending prospect of carbon regulation that did not exist so acutely in 2003, or other variables. In other words, there may be compelling reasons beyond corporate philanthropy for outside parties that would encourage their interest. Perhaps that is hopelessly naive on my part, but this is what we must fully explore and hopefully unlock.

Poston responded that he hadn't seen those compelling reasons. Although the potential return on investment was great in social terms, it was "non-existent in economic terms." Schwartz agreed with that conclusion, but argued that to come up with a viable Plan B, they needed to

> fundamentally alter our assumptions as we strive to come up with a new approach. So if we start the process with the goal of creating more self interest from the private sector (by granting more IP exclusivity, allowing vendors to compete, etc.), would that not change our thinking on how we might structure things? In other words, do veheicles [sic] like TIAs, loan guarantees, etc. become more viable tools if, at the outset, we seek to avoid a construct which is as "philanthropically" focused as the current deal seems to be?

Poston responded, "I am not certain how we can fundamentally alter the economics." He continued:

> The economics of our power production require other sources of revenue to offset the additional costs associated with carbon sequestration. . . . I have not heard of other revenue streams being identified except looking for participation from philanthropic organizations . . . but how would that play in the press? "DOE unable to support its own priorities; competes with the needy for funding?"

Schwartz admitted that "absent a basic change in some of the underlying assumptions, this is a circular exercise in which <u>we will always arrive at the rightful conclusion that the current arrangement is the best mechanism for achieving our goals</u>" (emphasis added).[97]

---

[96] E-mail entitled "Re: OGCC/CCS Process Integration Made Simple" from Thomas Sarkus to Jay Braitsch, Nov. 7, 2007.

While this discussion was going on, Poston also wrote of his strong misgivings to Thomas Brown, the director of the Office of Contract Management.

> Yesterday's meeting on what to do if an agreement on a revised Cooperative Agreement could not be reached included new participants but not new insights or conclusions.
>
> A very optimistic perspective was being offered on the possibilities of what we could do differently. I did try an [sic] add . . . an element of reality in that we took our best approach with the initial award and that unless we have changed our program needs (which we have not), have reduced our cost drivers (which we have not), or can introduce new money (which we might be able to but on a such a small scale that it is immaterial) I could not see much choice except to step back and focus the Department's efforts on R&D. . . .
>
> There are NO differences from 2003 so my response will sound like a broken record – if the current deal can not be satisfactorily restructured take our money and focus on R&D.[98]

But as Der told his staff: **"Doug [Schwartz] wants new ideas . . . . Doug is driving this with other hot shot project finance guys. . . . Have fun in this dump."**[99] (Emphasis added)

Operating on Dual Tracks

By the end of November, Sell was making daily requests for a detailed Plan B draft.[100] There is no indication that this option was ever shared with the Alliance until DOE made the announcement on December 18 that it was going to restructure FutureGen.

DOE's work on the Environmental Impact Statement required under the National Environmental Policy Act (NEPA) for the four finalist FutureGen sites was going forward as scheduled. DOE's October fact sheet on FutureGen mentioned that there were cost increases, but that they were "consistent with the increases seen in similar power plant projects and construction projects."[101] On October 30, a DOE employee said DOE was "diligently working"

---

[97] Series of e-mails entitled "RE: FutureGen Plan B" between Bradley Poston and Doug Schwartz, Nov. 6-9 and 15-19, 2007; undated memorandum entitled "Subject: FutureGen Option B" from Poston's files. Exactly what this change would be was unclear. In March of 2007, Thomas Shope testified before the House Energy and Commerce Committee that CCS technology would not be reliably available for commercial deployment until 2045 at the current level of funding for CCS and advanced power generation technology. George Rudins, then deputy assistant secretary for coal power systems, stated that the schedule could be accelerated by 20 years, but required annual federal funding of $1 billion plus deployment incentives. "It assumes a greatly expanded CCPI program and R and D. It also assumes a greatly expanded FutureGen program." E-mail entitled "Re: Date for CCS commercialization" from George Rudins to Frank Burke, March 7, 2007.
[98] E-mail entitled "FutureGen" from Bradley Poston to Thomas Brown, Nov. 7, 2007.
[99] E-mail entitled "This Coming Week" from Victor Der to Jarad Daniels, Nov. 9, 2007.
[100] E-mail entitled "RE: Fg" from Doug Schwartz to Andrew Patterson, Nov. 30, 2007.
[101] DOE, "FutureGen, FC26-06NT42073, October 2007, p. 3.

23

to complete the NEPA process and issue a Record of Decision (ROD) by the end of 2007.[102] The final EIS was issued on November 9.[103] On November 15, Albright and Slutz recommended that Secretary Bodman sign a letter to the Illinois Congressional delegation responding to an October 25 letter expressing concern about meeting the year-end deadline for a site selection. In that letter – which he later said was a mistake – Secretary Bodman repeated the commitment to complete the NEPA process and issue the ROD in a timeframe that supported FutureGen site selection by the end of December. Albright and Slutz also reminded the Secretary that the Texas legislature had passed incentives for a site in its state which would expire at the end of the year.[104] In late November, NETL staff was discussing a "big event" with DOE participation when the Alliance announced its final site selection.[105] By mid-December, sign-offs were being obtained on the ROD. The "potential" ROD signing was set for December 17 or 18, and a letter was drafted to the Alliance to that effect for Secretary Bodman [106]

At the same time, the Alliance also was pushing forward. In early December, it issued Secretary Bodman an invitation to the site selection announcement on December 17.

But the negotiations were not going well. On December 6, the Alliance sent a letter to Albright stating that it wanted to proceed under the existing cooperative agreement until "costs and risks can be properly assessed with input from the upcoming preliminary design report and cost estimate." The Alliance members did not want to accept considerably more financial risk without this information which "both parties previously agreed would be a precursor to these discussions." The Alliance also accused DOE of taking away the legal and financial options that would help it manage risk even though they had been available under other cooperative agreements, but assured DOE that its members would honor their obligations. The Alliance said both parties should "convey positive messages about the project" and not suggest that the current agreement was "anything less than a 'good deal.'" Assuming release by DOE of the ROD by December 17, the Alliance would make the site announcement on December 18.[107]

In a detailed attachment, Alliance CEO Mudd laid out the basis upon which the Alliance was originally formed:

1.   20 percent cost-sharing;

---

[102]E-mail entitled "Re: latest version" from Joseph Giove to Jarad Daniels, Oct. 30, 2007. A Record of Decision accepting the EIS must be signed by the agency before any federal funds can be expended.

[103] The final EIS was published in the *Federal Register* on Nov. 16. EIS No. 20070489, 72 *Fed.Reg.* 64619, Nov. 16, 2007.

[104] Letter from Michael Mudd to Secretary Bodman, Oct. 25, 2007; memorandum for the Secretary entitled "ACTION; RESPONSE TO LETTER FROM Illinois Congressional Delegation. At least two of the letters were signed, but not until Nov. 30. In a hearing before the Energy and Commerce Committee on Feb. 7, 2008, Secretary Bodman said it was a mistake. Letter dated Feb. 12, 2008, from Sen. Dick Durbin and Rep. Tim Johnson to Secretary Bodman.

[105] E-mail entitled "RE: SENSITIVE: FG Site Selection coordination????" from Thomas Sarkus to Victor Der, Carl Bauer and Miles Keith, Nov. 20, 2007.

[106] E-mail entitled "Cover Memo for FutureGen ROD," from Mark Matarrese to James Slutz, Victor Der, Jarad Daniels, Andrew Patterson, Kevin Graney, Raj Luhar, John Grasser and Robert Tuttle, Dec. 12, 2007; e-mail entitled "FG Draft Bodman Reply 11-15-07.doc" from Thomas Sarkus to Joseph Giove and Thomas Russial, Nov. 15, 2007.

[107] Letter from Michael Mudd to Bud Albright, Dec. 6, 2007.

2.   no repayment requirement from industry partner;

3.   ability to vest ownership of plant with industry partners;

4.   potential for program income to be shared among project participants;

5.   100 percent of post-project revenues to industry partners; and

6.   advanced appropriation of $300 million by DOE

But the Alliance members had given up many benefits by forming as a 501(3)(c) non-profit corporation, which meant that no income or proceeds could go back to the original members, but must be reinvested in public benefit research and development. They got no intellectual property rights. The cost share increased to 26 percent. There was an agreement to negotiate limits to the federal investment subject to escalation after there was a more detailed site-specific design and cost estimate. Mudd also pointed to the offers made by the Alliance to share revenues and to share proceeds from the sale with DOE.[108]

Slutz responded in a short letter stating that DOE was evaluating its "next actions" with respect to the Alliance and the FutureGen project. He further said that the Alliance had scheduled its final site selection announcement without consulting with DOE – although DOE had been aware for months of the plan to make the announcement by the end of the year – and that DOE would consider it "inadvisable" for the Alliance to do so because DOE did not anticipate issuing the ROD.[109]

"Sanity Check"

In early December, Brad Poston was asked for a last "sanity check" on Plan B. In a meeting with Andrew Patterson, a senior policy adviser, Poston said that the most critical question was whether industry would want to participate and reminded Patterson that four years ago, industry had shown little interest in FutureGen. "[W]e would be asking a utility stereotyped as risk averse [sic] organization, to use our unproven design on their $2.5B investment."[110]

DOE top officials weren't having any of it. On December 7, Albright told Jeff Kupfer, Secretary Bodman's chief of staff, that further negotiations with the Alliance were "at best fruitless and likely counter-productive." Albright had a new overall plan, but needed the approval of Sell, the Secretary and the White House.[111]

On December 11, DOE briefed the National Economic Council deputies on the new plan. Secretary Bodman briefed the NEC "principals" on December 14 on DOE's intent to restructure.[112] The "new strategy" was laid out in a briefing memorandum. He would cap the government's financial exposure and pointed to developments, such as tax credits and loan guarantees for clean coal projects, that had occurred since FutureGen was conceived in 2003. DOE would issue a competitive solicitation "aimed at accelerating near-term commercial deployment of integrated IGCC commercial power plants with cutting-edge CCS technology."

---

[108] *Ibid.*

[109] Letter from James Slutz to Michael Mudd, Dec. 11, 2007.

[110] E-mail entitled "RE: FutureGen" from Bradley Poston to Thomas Brown, Dec. 5, 2007.

[111] E-mail entitled "RE: FutureGen" from Bud Albright to Jeffrey Kupfer, Dec. 7, 2007.

[112] Attachment to e-mail entitled "FutureGen Timeline.doc: from Doug Schwartz to Kasdin Miller, Jan. 24, 2008.

DOE would fund only the CCS component of multiple IGCC plants, which it estimated would cost $350 - $500 million per plant. DOE's unnamed "experts" believed there would be "significant" private sector interest, although it had not discussed this with the private sector.[113]

Good Faith?

Whether DOE was operating in good faith during these negotiations with the Alliance is highly questionable. Secretary Bodman's intense dislike for the project was well-known by his staff. Undated notes recording a meeting about the legal obligations of the Department related to FutureGen read as follows: "S-1 [Bodman] aggravated by this project. Bob Card [former DOE undersecretary] deal. Trying to do everything in one project get smart on alternative options. Can we turn this off/redirect?"[114] At the end of September, Albright told FE "to work under the assumption that a threshold at the 1.8B figure with a 50/50 split afterwards, with some adjustment for increasing Alliance membership, would be sufficient."[115] But on October 25, an FE employee walked into a meeting with several high-level DOE officials, including Albright, Alexander (Andy) Karsner, the assistant secretary for energy efficiency and renewable energy, and Karen Harbert, the assistant secretary for policy and international affairs.

> The topic of discussion seemed to be how best to kill FutureGen. It was great fun, with Karsner leading the charge by suggesting that we just compete FutureGen under the loan guarantee program and let industry fight over who gets the Federal cost share, and touting how they make industry eat all the cost escalation in their biomass contracts.[116]

Interestingly, earlier in the year, Albright had been quoted as saying that any action on climate change had to involve the rest of the world. "Unless China and India are acting with us, it's pointless. They emit more carbon dioxide than we do."[117] Even though DOE and the Alliance had accomplished that goal and had both China and India as FutureGen partners, Albright was now in the lead to dismantle it.

Secretary Bodman appears to have made it clear to DOE staff that he did not care about the overarching goals of FutureGen, but only its cost. As Bradley Poston wrote in the midst of his efforts to contribute to a new plan, "I have an imperfect . . . understanding of the program; the current market conditions; and the changes in operating parameters from four years ago when the original acquisition strategy was developed. I see the true issue to be money and our ability to cap our financial exposure."[118]

Bodman's letter to Alliance CEO Mudd at the end of October stating that the ROD would be completed in time for a site announcement at the end of December appeared to be a commitment to the original FutureGen. But in December, Doug Schwartz, Albright's chief of

---

[113] *Ibid.*

[114] Undated, handwritten notes from the Department of Energy. Author not identified.

[115] E-mail entitled "FG – update" from Jarad Daniels to Thomas Russial, Sept. 26, 2007.

[116] Untitled e-mail from Jarad Daniels to Victor Der, Oct. 25, 2007.

[117] Biography of C.H. Albright Jr., *The Almanac of the Unelected, 2007,* Bernan Press, Lanham, MD, p. 140.

[118] E-mail entitled "RE: FutureGen Plan B" from Bradley Poston to Doug Schwartz, Nov. 9, 2007.

staff, said everyone was "conveniently forgetting" one thing: "[W]e're here b/c S-1 [Bodman] wants to kill FG as its [sic] currently contemplated, with or without a Plan B."[119] It was also clear that everyone knew that Plan B had a very good chance of failing to meet the original goals. It would be cheaper, but it might not work, and carbon capture would then be delayed. "We discussed the additional risk to the company building the plant and if they would actually be willing to take on this risk. I don't think we will know that until we put out a RFI and see what industry says," Sarah Magruder Lyle, DOE's White House liaison, wrote. The "message" focus would be on fiscal responsibility. There would be no fully funded advance appropriations for Plan B. Research would continue under the Clean Coal R&D program as in the past.[120]

It is also clear that the Alliance did not know the details of Plan B during the negotiations, although Albright may have discussed it generally with some of the member companies.[121]

The Decision

White House staff was expressing "much angst" over what Plan B would mean for commercial deployment of CCS technology.[122] DOE officials asked for a clear deployment timeline of "educated guesses and assumptions." The response was lukewarm at best even from the policy shop.

> Schedule for plan B is commercial scale operation of two or three plants by 2015 with the demo lasting until 2018. <u>One could argue that you would have commercially deployed plants in 2015 and at a minimum you can argue that you would have them at 2018 assuming that **they are still doing CCS after the demo**</u> (emphasis added).

On the other hand, FutureGen would operate from 2012-15. But if one "aggressively" assumed it would take three to five years before a commercial plant was built, you could claim the 2018-20 timeframe for the first commercial deployment – not exactly an acceleration from the original FutureGen.[123]

Nonetheless, the DOE higher ups had made their decision: Plan B would be rolled out with the promise that it would be better, faster and cheaper than the original FutureGen, regardless of the economics, industry interest, and the predictions of their own staff. Secretary Bodman communicated that to Senator Durbin in a phone call that apparently occurred on

---

[119] Untitled e-mail from Doug Schwartz to Julie Ruggiero, Dec. 10, 2007.

[120] E-mail entitled "Future Gen B Dec 12 2007 Final.doc" from Sarah Magruder to Karen Harbert, Dec. 12, 2007.

[121] E-mail entitled "RE: FutureGen Timeline.doc" from Mary Egger to David Hill, Jan. 24, 2009.

[122] E-mail entitled "timeline" from Jeffrey Kupfer to Bud Albright =, Doug Schwartz and James Slutz, Dec. 13, 2007.

[123] E-mail entitled "timeline" from Jeffrey Kupfer to Bud Albright, Doug Schwartz and Jim Slutz, Dec. 13, 2007; e-mail entitled "RE: timeline" from Andrew Patterson to Mr. Schwartz and Mr. Slutz, Dec. 13, 2007.

December 13.[124] On that same day, the NEC principals met and approved a restructuring of FutureGen if the Alliance didn't agree with all of DOE's demands.[125]

Victor Der, DOE's deputy assistant secretary for clean coal, was blunt in his opposition. Plan B was only a demonstration which "will likely use more conservative, more costly and substantially less efficient IGCC-CCS technologies rather than the more aggressive technologies being developed in our R+D program aimed at potential cost and energy penalty reductions. . . . Under plan B we would still have to follow up with sequential CCPI type demos which would incrementally add one or two advanced technologies at a time. This serial approach costs us time to fully deploy CCS globally." Der went on to say that his group's estimate that Plan B could delay by at least 10 years full commercial deployment of low-cost, low energy advanced CCS technology that could be transferred to developing countries wasn't included in the final analysis. A follow-up e-mail stated that affordable CCS technologies also would not be available in time for the expected turnover of the existing fleet of coal power plants in the U.S.[126] DOE officials responded by saying they were continuing to work "on a scenario that allows us to reduce/eliminate the 10 year deployment delay."[127]

Impact of OMB Budget Cuts

Secretary Bodman wasn't the only high-level government official not on board with the President's initiative. In September, DOE's budget shop told FE that the President's budget had additional funding that enabled FutureGen to stay on track and supported the baseline schedule. It reflected the ramp-up of activities as the program moved toward full-scale operation in 2012. FY 2009 activities included the complete detailed design of a prototype plant, money to initiate construction and the continued procurement of long-lead equipment.[128] But in November, the Office of Management and Budget (OMB), which was well aware of Bodman's opposition, eliminated all of the climate change funds from FE's budget.[129]

In early December, James Connaughton, the chairman of the President's Council on Environmental Quality (CEQ), met with representatives from Fossil Energy to discuss clean coal research in preparation for his attendance at the United Nations Framework Convention on Climate Change in Bali. Connaughton – who may not have been fully aware of the unrelenting drive toward Plan B –said that the U.S. had two options: either invest billions of dollars to develop the technologies to address climate change; or face a new regulatory environment that

---

[124] "Meeting Memorandum" to Secretary Bodman from Lisa Epifani regarding phone call to Senator Richard Durbin scheduled for December 13, 2007. Other reports put the call on December 14 and we know that it was postponed at least once from December 12. However, the call did occur.

[125] Attachment entitled "Purpose of Meeting" to e-mail entitled "FG principals mtg statement.doc: from Mary Egger to Mary Egger, Jan. 24, 2008.

[126] E-mail entitled "Re: timeline" from Vic Der to Mr. Slutz, Carl Bauer and Scott Klara, Dec. 13, 2007; "What 'Plan B' would NOT accomplish," attachment to e-mail entitled "FW" FG Plan B" from Jarad Daniels to James Slutz, Dec. 13, 2007.

[127] E-mail entitled "FW: FutureGen/CCPI funding (With brackets) from Darren Mollot to Jay Hoffman, Dec, 17, 2007.

[128] E-mail entitled "Proposed Change for FutureGen" from Karen Brown to Patty Graham, Robert Pafe, Jarad Daniels and Jordan Kislear, Sept. 28, 2007.

[129] E-mail entitled "Re" FY 2009 Budget intelligence" from Jeffrey Kupfer to Steve Isakowitz and Clay Sell, Nov. 15, 2007.

would not advance the technology. He also said that the U.S. needed to elicit more parallel activity in China and India.[130]

Connaughton's concerns were to no avail. On December 11, while he was in Bali, he received an e-mail from Karen Harbert at DOE. "I know how busy you are in Bali, but without significant interest by WH offices, we will not have a serious effort in climate," she wrote – and there was no such interest. Harbert went on to says that in the FY08 budget request, DOE had shifted over $500 million toward high-priority programs, including Futuregen, in clean coal and nuclear research and development, but OMB had eliminated all of the additions. Harbert acknowledged that the heavy emphasis on CCS would also help reduce emissions in China and India, but that OMB had eliminated "all funded increase for clean coal, greatly undermining plans for critical demonstrations as well as FutureGen."[131] In a related e-mail, Connaughton was portrayed as being

> very apprehensive about the international piece – and how we deal. What happens to other countries, etc. Bottom line is that he likes his international talking point and wants to keep it. CEQ is going to try to set up a call for you [Harbert] and him sometime later today – so that you can convince him that this is meangeale [sic]. Hopefully you can do that.[132]

These budget cuts made it extremely difficult, if not impossible, to build the original FutureGen under any circumstances, as the DOE expenditures were front-loaded in the project schedule, even with a 50/50 cost share after the initial $1.8 billion was spent.

Announcement by Alliance of Final Site Selection

The Alliance's time line established the end of 2007 for the announcement of the final site decision. As DOE had completed the final EIS, the Alliance scheduled the announcement for December 18. The winner was the State of Illinois with a site near the city of Mattoon. But within hours, DOE, in a statement made by James Slutz, said that "the public interest mandates that FutureGen deliver the greatest possible technological benefits in the most cost-efficient manner. This will require restructuring FutureGen to maximize the role of private sector innovation, facilitate the most productive public-private partnership, and prevent further cost escalation."[133] DOE also stated that it would not sign the Record of Decision on the EIS which was required before any federal project construction funds could be expended.[134]

---

[130] E-mail entitled "Recap of CEQ meeting on FY09 Passback" from Jarad Daniels to Victor Der, Nov. 30, 2007. CEQ did host a meeting on FutureGen in early October to which representatives from the White House, the Office of Science and Technology Policy and DOE were invited. E-mail entitled "CES mtg. re. FutureGen" from Doug Schwartz to Nell Kinsey, Oct. 2, 2007.

[131] E-mail from Karen Harbert to John Herrmann, NSC, Dec. 11, 2007, enclosing e-mail entitled "DOE Appeal Status" from Ms. Harbert to James Connaughton, Dec. 11, 2007. Harbert said DOE had appealed $380 million but recovered only $24 million.

[132] E-mail from Jeff Kupfer to Karen Harbert, undated.

[133] "Statement from U.S. Department of Energy Acting Principal Deputy Assistant Secretary for Fossil Energy," Dec. 18, 2007.

[134] AP, "Mattoon, Ill. picked for FutureGen pollution-free coal plant," Dec. 18, 2007; e-mail entitled "Backlash draft" from Julie Ruggiero to Megan Barnett, Dec. 18, 2007.

Plan B Goes Forward

During January, there were some continued negotiations with the Alliance as the White House had not yet officially signed off on Plan B. On January 10, the Alliance sent a letter proposing a "new approach to financing FutureGen." It would increase its cost share if overall costs went up, make post-project repayments and do partial bank construction financing. Under this approach, the Alliance claimed the final taxpayer investment would be no greater than it was on the day President Bush announced the project.[135] But DOE internally remained focused on Plan B. Albright told DOE and White House staff that "[r]egardless of the value of their proposal, we need to continue to move expeditiously with the new direction rollout." The Alliance, for its part, refused to share the details of its proposal unless there was an "in person" meeting.[136] DOE's clean coal staff had one job left: make the fantasy that was Plan B look good on paper.

Putting together a seemingly logical story around Plan B to sell to the White House, Congress, the press and the public was not an easy job. After reviewing a rough outline of the program plan, Victor Der forwarded it to Jay Hoffman, DOE's director of program analysis and evaluation with this message: "Here's the Frankenstein. I'll be calling NETL to see where they are in the electrodes development to make it walk." [137] Hoffman responded with a new "FutureGen Plan B Storyline." The main rationale, according to Hoffman, was "a more appropriate public/private cost allocation between DOE and industry. Secondary benefits may include accelerated commercial demonstration and more carbon-free power, but these are not driving reasons for why Plan B is being developed" (emphasis added), Hoffman wrote. IGCC technology was "a largely commercially proven technology" and didn't need government assistance. CCS, on the other hand, was "largely unproven," and DOE would pay for the resulting research and development, operating and maintenance and parasitic energy losses that the private company would incur.[138]

After looking at the "story line," Der wasn't convinced. "[T]he FrankenGen document, I mean, New FutureGen, needs to be taught to walk first, before it can hop on a Harley."[139]

A few days later, Secretary Bodman was briefed by Albright on DOE's "new focus." The possible "secondary benefits" became real benefits in this presentation. Because of construction costs, "growing near-term interest in carbon dioxide regulations and states beginning to require CCS or the flexibility to add CCS for siting/permitting of coal plants," DOE was now going to focus on "first-of-a kind full utility-scale demonstrations and developing data on commercial cost, integrated IGCC-CCS performance and reliability to reduce risk, confirm economics and facilitate industry-wide private capital offerings." This would allow for early deployment of

---

[135] Letter dated Jan. 10, 2008, from Michael Mudd to C.H. Albright, p. 1

[136] E-mail entitled "RE: FutureGen" from Bud Albright to Cynthia Bergman, Charles Blahous, Jeffrey Kupfer, Andrew Beck and Lisa Epifani, Jan 16, 2008.

[137] E-mail entitled "Plan B Program Plan 12_20_2007.doc" from Victor Der to Jay Hoffman, Jan. 2, 2008.

[138] E-mail entitled "FY09 FutureGen Program Plan Storyline" from Jay Hoffman to Victor Der, Jan. 4, 2008.

[139] E-mail entitled "FW: A Program Plan for Demonstration of Integrated Electric Power Production and Carbon Sequestration" from Victor Der to Jay Hoffman, Jan. 2, 2008; e-mail entitled "RE: FY09 FutureGen Program Plan Storyline" from Victor Der to Jay Hoffman, Jan. 4, 2008.

"nearer term IGCC-CCS technologies" at commercial plants and would also address the "very critical technical feasibility question" of a near-zero emission coal plant. There would be a minimum of two 600 MW plants, each of which would capture and store at least 1 million metric tons of CO_2 per year. Staff did note, however, that cost reductions and competitive technology were still needed for full deployment, and that those technologies would still have to be demonstrated later. There was no explanation about why industry would test technology that was not yet cost-effective.[140]

The Department also struggled to put together an internal "strategic plan" for the White House that would incorporate – with some facial credibility – the new FutureGen structure while claiming to maintain the original goals of an IGCC, near-zero emission plant. DOE postulated that because of the challenges of getting coal-fired plants licensed, this "change in the market landscape" had "catalyzed the need" to demonstrate the commercial viability of an IGCC/CCS plant. However, because of the uncertainty about the cost and performance of such plants, plans for them were being abandoned or postponed. "Unless the production of electricity from coal integrated with sequestering carbon dioxide can be shown to be commercially feasible and cost competitive, the coal industry will not make the investments necessary to fully realize the potential energy security and economic benefits of this plentiful, domestic energy."[141] Reducing that uncertainty of course, was exactly what the original FutureGen was supposed to demonstrate. But in an inexplicable shift in reasoning, DOE then said that it would achieve its goals more quickly if it could attach a CCS technology to a commercially built IGCC plant. It would speed up commercialization, help drive the regulatory framework and address the "very critical technical feasibility question of advanced technology clean coal plants."[142]

FE did not go down without a fight. On January 10, Jay Hoffman, director of the Office of Program Analysis and Evaluation, who was working on the FY 2009 budget, laid down the law to Victor Der and Jarad Daniels.

> Let me get right to the point. As written, the CFO's [chief financial officer] office will not concur on the project plan. It is sorely lacking in detail and analysis, and provides little defense or answer to the difficult questions we will field from the WH, the alliance, and ultimately the public/Congress . . . . My expectation was for your office to develop a solid, analytically supported plan that at a minimum included the suggested analysis, with the caveat that you could determine how best to frame the story around that analysis.

Hoffman said he expected a revised project plan for the decision makers that would be "bullet proof and ready for the WH." It needed to describe what went wrong with the original FutureGen and why Plan B would be successful, including why industry would buy into it.[143]

---

[140] "New FutureGen: Briefing to Secretary of Energy." Jan. 9,2008, pp. 2-3.
[141] "Draft Strategic Planning Document for Revised FutureGen: Demonstration of Integrated Electric Power Production and Carbon Capture and Sequestration," December 2007, p. 4.
[142] *Ibid.*, p. 2.
[143] E-mail entitled "FW: FY09 FutureGen Program Plan Storyline" from Jay Hoffman to Victor Der and Jarad Daniels, Jan. 10, 2008.

The goals listed in the new FutureGen in the final drafts read like DOE's ultimate coal dream: it would validate CCS at multiple sites, it would inject and monitor CO_2 at multiple geologic formations, integrate CCS with multiple gasification-based power production technologies; develop a regulatory and permitting system; provide the possibility of international participation at more than one project; produce a more comprehensive and reliable set of operating data, and promote early widespread deployment of IGCC-CCS technology. In addition, it would capture at least 90 percent of CO_2 and mercury, 99 percent of sulfur, and reduce NOx and particulate emissions. And all this came with a lower federal price tag.[144]

There, of course, was one big problem: Plan B would cost the power generator a great deal of money in capital, operating and maintenance and parasitic energy loss costs. DOE's program and budget people struggled for a month to put together a cost estimate that would be lower than the original FutureGen. Initially, DOE was going to pay for the parasitic energy loss, but that became too expensive so it was deleted. The government would only pay capital costs for the CCS addition to an IGCC plant. Questions raised about the readiness and costs of the CCS technology were ignored. "Biggest area of concern remain 'new technology' and the insertion of this new technology into a 'generic' plant; not sure of the true impact and cost implications," the director of the Office of Engineering and Construction Management wrote.[145] "Taking these concerns in totality, and looking at it from industry's perspective, how does this uncertainty impact the profit potential of the project? At the end of the day, this will determine participation by industry," other DOE officials warned.[146]

There was another concern: the White House hadn't yet signed off on DOE's plan.[147] The final White House meeting was on January 25. DOE presented a strategic plan, complete with proposed press release and request for information (RFI), for Plan B to go out on January 31. DOE would contact the Alliance and make a final offer: the Alliance had until January 29 to accept the terms, which had a "50/50 cost share after the 1.8, and stating that the Alliance contribution may not include project financed debt." If the Alliance did not accept those terms, DOE would announce its new approach and put out the RFI on Jan 31.[148]

The White House meeting was to be hosted by Keith Hennessey, NEC's director and economic adviser to President Bush. Invited participants included OMB Director Jim Nussle; David Addington, Vice President Cheney's counsel; Press Secretary Dana Perino; Joel Kaplan, White House deputy chief of staff; CEQ Chairman Connaughton; Presidential Counselor Ed Gillespie; Charles Blahous, NEC deputy director; and Dr. John Marburger, director of the Office

[144] "Draft Strategic Planning Document, December 2007, *supra,* pp. 3-4.

[145] E-mail entitled "RE: Cost estimates for FutureGen Plan B" from Paul Bosco to Jay Hoffman and Melvin Frank, Dec. 19, 2007.

[146] Attachment to e-mail entitled "plan b observations.doc" from Jay Hoffman to Andrew Patterson and James Slutz, Dec. 13, 2007. Also, the director of DOE's Office of NEPA Policy and Compliance didn't think that DOE had a credible NEPA strategy for Plan B since only one of two units at a site would capture 90 percent of the CO_2, and there were other pollutants. FutureGen was a "major source" under the Clean Air Act, she reminded the general counsel's office. E-mail entitled "re:: fg DOCUMENTS" FROM Carol Borgstrom to Mary Egger, Jan. 16, 2008.

[147] E-mail entitled "draft talking points for S-2 tomorrow with Texas Railroad Commission" from Jarad Daniels to Kevin Graney, Jan. 17, 2008; e-mail entitled "Re: FutureGen – Ltr to Alliance (jan 18).doc" from Adam Ingols to Doug Schwartz, Mary Egger, James Slutz and Eric Nicoll, Jan. 18, 2008.

[148] E-mail entitled "Re: FutureGen issues and actions" from Scott Klara to Jarad Daniels, Jan. 24, 2008.

of Science and Technology Policy.[149] Sell and Albright were to "tell WH details of going forward and get blessing."[150]

Albright and Sell told the NEC principals everything they needed to hear to believe that the Bush initiative would remain intact. The restructured FutureGen would achieve all of the primary technical goals of the original project which was "no longer optimal to achieve the goal of accelerating the commercial demonstration and deployment of advanced, integrated coal-based power systems including CCS." But the government's financial exposure would be limited to mitigating the "incremental risk of the addition of CCS" while its investment would be leveraged "across a wider range of nearer-term coal based IGCC-CCS projects."[151] Not only would it accelerate deployment of CCS technology, restructured FutureGen would establish the technical feasibility and economic viability of producing electricity and hydrogen from coal with near-zero emissions. It would verify the sustained, integrated operation and effectiveness, safety and permanence of a coal conversion system with carbon sequestration, it would establish standardized technologies and protocols for CO_2 monitoring, mitigation and verification, it would sequester at least 1 million tons of CO_2 in saline formations; it would capture at least 90 percent of the CO_2 emitted; 90 percent of the mercury emitted; 99 percent of the sulfur and high levels of NOx and particulate emissions. There would be a more rapid investment by industry in multiple demonstrations of "near-commercially available technologies" for CCS.[152]

Additionally, because of the loss of the "living laboratory" element of FutureGen, there would be a "fresh look at the commercialization profile of key FE technologies." This was a particularly puzzling statement because the table of technologies that followed made it clear that most of them were still at the bench or laboratory stage of development, and FE would have to find alternative host sites. There were other confusing statements. While admitting that Plan B would delay the cost-reduction improvements that were ultimately needed for coal/CCS plants to be an attractive commercial option in both the U.S. and internationally,[153] Sell and Albright claimed that it would demonstrate "commercial feasibility." Private companies apparently were now expected to quantify the technical and economic risk associated with near-zero emissions coal plants, thus "enabling private financing decisions of future plants of this type" and facilitating "industry-wide private capital offerings."[154]

But deep in the strategic plan was the recognition that incorporating CCS on a commercial-scale IGCC plant added capital and operating costs and "is still perceived by the electricity generation industry as an emerging technology. Concerns remain over the integration and scale-up risks associated with IGCC, and a cost gap still remains when compared to conventional coal power plants." Industry's reaction to the new program would depend on the

---

[149] E-mail entitled "1/24 FutureGen Principals Meeting – TIME CHANGE" from Kristin Marshall to Ann Merchant et al., Jan. 23, 2008.

[150] E-mail entitled "Re: FutureGen issues and action" from Scott Klara to Jarad Daniels, Jan. 24, 2008.

[151] "Draft Strategic Planning Document for Revised FutureGen: Demonstration of Integrated Electric Power Production and Carbon Capture and Sequestration," Jan. 30, 2008, pp. 2-3 and 8.

[152] *Ibid.,* p. 3.

[153] "Under Revised FutureGen commercial deployment of cost-reduction improvements could be delayed unless other test approaches are found, such as designing limited test capability . . . into Revised FutureGen and CCPI demonstrations." *Ibid.,* p. 7.

[154] *Ibid.* pp. 3 and 7-8.

"magnitude of the government's commitment to the project" and its ability to "reasonably satisfy" those concerns and allow the plants to function competitively. And, of course, there was that troubling issue of liability for the sequestration of CO2.[155]

DOE also claimed that its international partners would favorably respond, even though they no longer could share in the technology development or work at the new sites. Inexplicably, DOE found that the new approach would actually "raise the efficiency of information sharing."[156]

Albright and Sell were successful. By January 28, everyone in the White House was "on board" with the announcement for a restructured FutureGen.[157]

In the final strategic plan, DOE ignored every concern of its own staff. "Today, more than ever, the concept of FutureGen is a centerpiece for the future of coal utilization," the plan trumpeted.

> FutureGen directly addresses a primary goal of the Department of Energy's (DOE) 2006 Strategic Plan under the Theme for Energy Security to promote America's energy security through reliable, clean, and affordable energy: Environmental Impact of Energy: "Improve the quality of the environment by reducing greenhouse gas emissions and environmental impacts to land, water and air from energy production and use."[158]

January 30, 2008 announcement

Secretary Bodman met with the Illinois delegation on January 29 to forewarn them of the announcement. His plan was very poorly received by both Republicans and Democrats, who called it "unfair," "cruel" and "incompetent management." They asked how DOE could throw away Illinois' five years of work.[159] Just before the announcement, Illinois Republican Congressmen Tim Johnson and John Shimkus made an appeal directly to President Bush to save the project. The President said he stood by Bodman's decision.[160]

DOE then announced that it would "join industry" in its efforts to build IGCC plants by providing funding for the addition of CCS technology to multiple plants that would be operational by 2015. According to DOE, this would double the amount of CO2 sequestered

---

[155] *Ibid.*, pp. 16-17.

[156] *Ibid.*, p. 17.

[157] E-mail entitled "FutureGen" from Cynthia Bergman to Megan Barnett, Jan. 28, 2008.

[158] DOE, "Draft Strategic Planning Document for Revised FutureGen: Demonstration of Integrated Electric Power Production and Carbon Capture and Sequestration," Jan. 31, 2008.

[159] E-mail entitled "re: FG REDLIGHT – S-1 agreed to wait one day" from Jeffrey Kupfer to Eric Nicoll et al., Jan. 29, 2008.

[160] "Durbin sees 'uphill struggle' to save FutureGen; Energy Dept. confirms it is pulling its back for the coal-fueled experimental power plant in Mattoon, Ill.," *St. Louis Post-Dispatch,* Jan. 31, 2008, D2.

compared to the original FutureGen.[161] The restructured approach allowed DOE to "maximize the role of private sector innovation, provide a ceiling on federal contributions, and accelerate the Administration's goal of increasing the use of clean energy technology to help meet the steadily growing demand for energy while also mitigating greenhouse gas emissions."[162] Secretary Bodman also claimed that engagement with the international community would remain "an integral part" of DOE's efforts, although he had already been told that private companies would not be interested in freely sharing their technology with other parties, foreign or domestic.[163]

The mysterious "technology advance" that Secretary Bodman and others kept referring to was that, unlike in 2003, there were now over 33 IGCC plants that have been proposed, even though a number of them had already been cancelled. In a follow-up conference call with reporters, Sell claimed that "[t]his fact, this changing underlying market dynamic, underpins why we believe our new approach is fundamentally better to advance the state of carbon capture and sequestration." He expressed his confidence that restructured FutureGen was a better way to go. "We are making this project better and we are increasing substantially the likelihood of success."[164] Sell even claimed that the National Energy and Technology Lab's (NETL) work gave him that confidence, despite the fact that NETL, FE and others had been protesting for months that the new approach would not work.[165]

There was no discussion of who would take on the liability for sequestration or who was going to pay for the energy loss associated with CCS or how the technology had suddenly advanced to viable commercialization. DOE would issue a Request for Information to the industry to determine its views (which had not been sought before the announcement). It would be followed by a competitive Funding Opportunity Announcement.[166] Any loss of the research and development aspects of FutureGen would be made up in a significant increase in the FY 2009 clean coal budget.[167]

The RFI asked for input and public comment on the restructured FutureGen and expressions of interest from power producers who would consider participating in the revised initiative. These responses would help shape a competitive funding opportunity announcement expected to be released in June of 2008. DOE stated it was interested in funding multiple demonstrations of CCS technology at a commercial scale of at least 300 gross MW per unit plant power train per demonstration. It would contribute no more than the incremental cost of the CCS for one train. At least 1 million metric tons of CO_2 would be stored in a saline storage

[161] FE staff had told the policy and press staff that if they were going to maintain the 90 percent carbon capture goal, IGCC was the only credible approach. E-mail entitled "RE: FOR YOUR REVIEW – updated fact sheet and press release" from Jarad Daniels to Megan Barnett, Jan. 22, 2008.

[162] "DOE Announces Restructured FutureGen Approach to Demonstrate CCS Technology at Multiple Clean Coal Plants," press release, Jan. 30, 2008.

[163] *Ibid.*

[164] Transcript of Department of Energy conference call," Jan. 30, 2008. The speakers were Clay Sell and Secretary Bodman.

[165] *Ibid.*

[166] DOE press release, "DOE Announces Restructure FutureGen Approach to Demonstrate CCS Technology at Multiple Clean Coal Plants," Jan. 30, 2008.

[167] "FutureGen Talking Points", undated.

formation, and all emissions levels for other pollutants would meet the original FutureGen goals. Commercial operations were expected to begin in 2015.[168]

Response to Restructured FutureGen and Request for Information

The response was quick and skeptical with most of the media viewing FutureGen as dead. "The administration has long trumpeted technology, not regulation, as the answer [to global warming]. There was no trumpeting last week when it unexpectedly canceled FutureGen – its much-touted, $1.8 billion attempt to develop a cutting edge coal plant that would turn coal to gas, strip out and store underground the carbon dioxide that contributes to climate change, and then burn the remaining gas to produce hydrogen and electricity," the *New York Times* wrote. "And what of Mr. Bush's hydrogen-powered Freedom Car? That, too, has receded from view." The newspaper described the decision as ending a four-year-old program that had been described as "one of the boldest steps our nation has taken toward a pollution-free energy future."[169] The *St. Louis Post-Dispatch* opined that Secretary Bodman apparently missed the part of Bush's 2008 State of the Union address on the previous day where the President urged Congress to "fund new technologies that can generate coal power while capturing carbon emissions." *IEEE Spectrum* described the decision as bringing FutureGen to a "screeching end."[170]

The responses received in March from industry to the Request for Information were more damning. There were 49 responses, almost all of which took major "exceptions to the RFI specifications and near zero emissions objectives," a DOE summary document reported. Industry wanted the solicitation expanded to non-IGCC technology; a "substantial relaxation" of the 90 percent carbon capture requirement; government liability protection of the CCS aspects of the projects; elimination of the mandate to sequester 1 million tons of $CO2$ in a saline aquifer and permission to sell $CO2$ for enhanced oil recovery; guaranteed funding up front; an expedited NEPA process; a sharing of the additional operating and parasitic energy costs; and reductions in the performance targets of sulfur, nitrogen oxide, particulate matter and mercury. The comments also suggested that the schedule was unrealistic.[171]

The comments from the Coal Utilization Research Council (CURC), an industry advocacy group that focuses on the technology development steps necessary to achieve near zero emissions from coal power generation (and which opposed the termination of FutureGen), were particularly negative. There wasn't enough money for "multiple" CCS projects (CURC estimated at least $600 million needed for each project), nor was there any assurance that Congress would provide funding; 90 percent $CO2$ capture was not realistic for a commercial project; and non-IGCC projects should be considered.

---

[168] DOE, "Request for Information (RFI) on the Department of Energy's Plan to Restructure FutureGen," Jan. 31, 2008.

[169] "Higher Costs Cited as U.S. Shuts Down Coal Project," *The New York Times,* Jan. 31, 2008, C5; "Late and Lame on Warming," *The New York Times*, editorial, Feb. 4, 2008.

[170] "Back to the FutureGen," *St. Louis Post-Dispatch,* Jan. 31, 2008, C8; "U.S. Government Terminates Its Major Clean Coal Project," *IEEE Spectrum OnLine,*
http://blogs.spectrum.ieee.org/tech_talk/2008/02/us_govt_terminates_its_m.html

[171] DOE' "Expanded Summary of Comments Received Under DOE's Request for Information (RFI) on Plan to Restructure FutureGen," March 20, 2008.

Given the immature state of experience in using capture technology integrated with an IGCC, for example, CURC believes it is much more prudent to simply encourage the installation of CCS technology on a unit that will be commercially-operated rather than dictate the level of capture. Industry should be free to determine what level of capture of CO2 makes the greatest sense from both a cost and acceptable risk exposure perspective.

CURC also estimated that installing CCS systems on to commercial projects would cost hundreds of millions, if not billions, of dollars, and the owners "should not be restricted to the 90% capture requirement that is otherwise germane to a technology demonstration project (i.e. FutureGen)." Additionally, a much larger initiative was necessary to continue a large-scale, industry-supported CCS implementation partnership.[172]

These were the same points DOE staff had raised earlier. In an issues document based on the comments, DOE staff wrote: "In the current environment, utilities planning new base load power capacity have compelling incentives to adopt a 'wait and see' approach while issues related to retail competition and carbon management are resolved. Moving forward with CCS at this time, absent legislation or other incentives, would be imprudent." Industry also was expressing skepticism about government support for the new program because of the change in direction and the change in administrations.[173]

DOE plowed forward, reiterating once again to Illinois Congressional members that its approach would help permit new commercial coal plants.[174] However, it hid the supposedly "public" comments from the public and the press by refusing all requests to release them.[175]

But there were other public forums which clearly exposed the problem DOE was going to have in getting responsive proposals. In May of 2008, the greenhouse gas research and development program and the clean coal center of the International Energy Agency held a workshop on financing CCS. The workshop participants' view was that private investment in CCS in North America was an "unattractive financial option without Government incentives and a legal framework in place." As a representative of JP Morgan Chase said, CCS has no positive purpose. It only has a negative purpose to avoid the cost of putting CO2 into the atmosphere, and that has no cost in the United States. The investment banks wanted a "secure return on their investment, such as loan guarantees or tax credits." Legal and environmental liability was an

---

[172] "Comments Submitted to the Department of Energy by the Coal Utilization Research Council (CURC) in Response to a Request for Information (RFI) Issued by the DOE," March 3, 2008, pp. 1-3 and 4.

[173] DOE, "Revised FutureGen Project – Outstanding Legal, Contractual and Policy Issues," March 25, 2008, Rev. 1, p. 1. DOE also expressed the fear that if the CCS technology failed, because of the numerous plant modifications necessary in an IGCC plant to capture and sequester CO2, "the entire plant could be considered a stranded asset." Therefore, the entire cost of the plant could be included in the base for cost-sharing, as it had been in other projects were novel technology is being tested. *Ibid.*

[174] Letter to Rep. Tim Johnson et al. from Secretary Bodman, attached to "Memorandum for the Secretary" from C.H. Albright, Jr., to James Slutz, April 9, 2008.

[175] Despite requests under the Freedom of Information Act, DOE refused to release these comments or those submitted on the draft Funding Opportunity Announcement until this Committee requested them. It provided no legitimate reason for withholding the comments beyond a claim that there was proprietary information in some of the responses. See, e.g., e-mail entitled "FG docs" from Andrew Patterson to Scott Shiller, Victor Der and James Slutz, March 31, 2008.

issue, and insurance companies were not ready to take on this risk. Until there was greater regulatory and cost recovery certainty, the private sector would not invest. And, "ultimately, the willingness of ratepayers to pay higher electricity bills to pay for CCS, as reflected in decisions by local public utilities, will be critical to the financing of such projects," the participants agreed. "It is clear that CCS is not economic and subsidies will be needed for the first plants. . . . [F]inancing is the key and ultimately without financing there will be no CCS deployment."[176]

Funding Opportunity Announcement (FOA)

The Draft Funding Opportunity Announcement was issued on May 7, 2008. Despite the RFI comments, it remained focused on a gasifier technology. As CURC stated in its comments, the FOA described a commercial-scale project which included the goals and objectives of the original FutureGen, which was a publicly co-funded demonstration-scale project, and that was not viable.

> Included among our suggested modifications are changes to FOA requirements related to emission controls of criteria pollutants, beyond that which is required for permitting plants today, a level of CO_2 capture percentage that has not been previously achieved in power plants at a commercial scale, dates for operation that may be difficult to achieve and other criteria that also may not be realistic or prudent when measured against the business requirements of a facility, or facilities, planned and constructed to operate successfully in commerce.

CURC reminded DOE of its earlier comment that there was not enough money for multiple projects, and, since future funding was not guaranteed, "there are not clear reasons why an owner or operator can have confidence that the bulk of the funding for a selected project will be forthcoming at a later date." CURC recommended a reduction below the 81 percent CO_2 capture level, which it described as "not a reasonable approach" at this stage of technology development or integration. "Industry needs to obtain baseline data, demonstrated reliability and widespread confidence in CCS systems and these goals can be achieved more cost-effectively by requiring less aggressive percentages of capture."[177]

CURC also wanted more flexibility in the CO_2 storage site, a regulatory structure for CO_2 transport, a resolution of long-term liability issues, more favorable cost-sharing arrangements, including recognizing the parasitic energy loss as a cost, and modifications that made it clear that non-IGCC plants were eligible.[178] In a summary of the unreleased "public" FOA comments, DOE indicated that they were similar to those submitted by CURC.[179]

---

[176] IEA Greenhouse Gas R&D Programme, World Coal Institute and IEA Clean Coal Centre, "Summary Report on Expert Workshop on Financing Carbon Capture and Storage (CCS): Barriers and Solutions," May 28-29, 2008, pp. 2-3 and 8.

[177] CURC, "Comments related to the Department of Energy draft announcement #DE-PS26-08T00496 related to "RESTRUCTURED FUTUREGEN," May 21, 2008, pp. 2-3.

[178] *Ibid.*, pp. 6-7

[179] "DIFFERENCES BETWEEN RESTRUCTURED FUTUREGEN "DRAFT" AND "FINAL" FUNDING OPPORTUNITY ANNOUNCEMENT (FOA)," attached to e-mail entitled "FG Q&As for Final FOA.6-23-08.v4.doc" from Jarad Daniels to Keith Miles and Thomas Sarkus, June 23, 3008.

The final FOA made some of those changes. A non-gasification project did not have to produce at least 250 MW net electricity output but could be at a "commercially viable size." There was no mandatory ceiling on the project cost. The applicants must "propose" start-up by Dec. 31, 2015, but apparently had no obligation to meet that date. The demonstrations were "expected" to operate for 3-5 years and capture 1 million metric tons of CO_2 per year that would be put in a saline "formation," not an aquifer as originally required. There was no obligation to operate after the demonstration period, and monitoring of the sequestration site would continue for only two years after the demonstration was completed. DOE would contribute the lesser of (1) the incremental cost of implementing CCS on the demonstration unit; or (2) 50 percent of the total allowable project cost. DOE's maximum cost would be negotiated prior to the award. Applications were due on October 8, 2008, with selections made by the end of the year. [180]

In the final FOA, DOE bragged again that "[t]oday, more than ever, the FutureGen concept holds great promise for sustaining near-term coal utilization."[181] Internally, staff saw it quite differently. The goals that Secretary Bodman had promised when he rolled out the restructured FutureGen were no longer mandatory. "The reality of Financial Assistance awards is that they should be viewed as "best effort," Keith Miles wrote.

> DOE asks for the Applicant to address all of the requirements (goals and objectives), provide a Statement of Project Objectives (SOPO) as well as the evaluation criteria in the FOA, which will ultimately be reviewed by DOE with selections made. Unfortunately there are no "consequences" if they don't achieve the goals and objectives contained in their SOPO. DOE's only recourse is when an issue of "noncompliance" arises, or research misconduct.[182]

No one – except those who may have drunk the Kool-Aid at DOE – was surprised at the anemic response to the FOA. In the end, almost no one came to DOE's party, and it wasn't the party that had been advertised in the invitation. There were four applications, two of which did not come close to meeting the criteria. Neither of the survivors proposed an IGCC/CCS plant, but hoped to test out experimental carbon capture technology on existing facilities. It was reported that even those applications were incomplete.[183] In January of 2009, Secretary Bodman and his deputies slipped out of town minus viable projects or even press releases claiming success.

Relationship with International Partners

Despite the years-long push to get other countries involved in FutureGen and the emphasis by high-level Bush officials on international participation in FutureGen, DOE did not discuss its change in plans with its international partners. Nor did it take any steps to inform the State Department's and its own international staff, which were continuing to solicit foreign partners. In a presentation to Brazil in October, FutureGen was described as a "unique

[180] DOE, "Funding Assistance Funding Opportunity Announcement," June 24, 2008.
[181] *Ibid.,* p. 6.
[182] E-mail entitled "RE: Restructured FutureGen @ REMINDER COMMENTS DUE BY 10:30 AM" from Keith Miles to Jay Hoffman and David Pepson. June 23, 2008.
[183] "New Life for Clean Coal Project," *The Washington Post,* March 6, 2009, A1.

opportunity to prove carbon sequestration . . . [and] to advance IGCC technology." International participation would facilitate implementation of CCS in emerging economies.[184] In November, Secretary Bodman, who had met previously with Polish officials, sent a letter encouraging Poland to join the initiative.[185]

In December, Treasury Secretary Paulson in a speech before the Asia Society prior to another SED meeting with China stated that the FutureGen clean coal development partnership with China represented one "of the best areas of on-going cooperation."[186]

When Karen Harbert, DOE's assistant secretary for policy and international affairs, asked how international partners could be incorporated into the new FutureGen, she was bluntly told that it had no international component.[187] But when Japanese officials sent a draft of a "framework" for a FutureGen agreement between the U.S. and Japan and a $10 million contribution on January 18, Harbert told them to "hold tight." Japan had hoped to have it signed in the next week at the World Economic Forum and had already put $700,000 in its budget for the project.[188] In the final draft of the supporting documentation for the restructured FutureGen, DOE removed all references to foreign governments' having access to test demonstration results because "they wouldn't have access to any of the 'good' proprietary information, but rather only the non-proprietary information which DOE always makes publicly available for any of projects anyhow."[189]

In a draft memo prepared for James Slutz to issue after the January 30 announcement, the partners were to be told, "The commercial market place will be the mechanism to deploy new technology such as Integrated Gasification Combined Cycle (IGCC) with CCS." DOE was, however, "committed to an international outreach component" which was "critical to garnering broad acceptance of the new technology and fostering the replication of the near zero-emissions on a broad scale." In other words, "thanks, but no thanks."[190]

On Feb. 1, 2008, Secretary Bodman sent out letters to all the current and potential foreign partners telling them that FutureGen was being restructured to emphasize commercial demonstration of CCS with IGCC plants, and that he looked forward to "continued outreach" to the interested countries.[191] The first – and most angry – response came from Korea. Kijune Kim of the Ministry of Commerce, Industry and Energy, wrote,

---

[184] DOE, "FutureGen: A Path to Success: The Right Project at the Right Time," Oct. 17, 2007.

[185] Letter from Samuel Bodman to Piotr Naimski.

[186] Remarks by Secretary Henry M. Paulson, Jr. on "Maintaining Forward Momentum in U.S.-China Economic Relations," Treasury Department press release, Dec. 5, 2007, p. 2.

[187] E-mail entitled "RE: Int'l aspects of new futuregen construct" from James Slutz to Karen Harbert, Dec. 12, 2007.

[188] E-mail entitled "FW: Signature for the Framework on FG Project between DOE and METI etc." from Jarad Daniels to Joseph Giove, Jan. 18, 2008; e-mail entitled "Re: FutureGen Framework Agreement" from Talashi Naruse to Joseph Giove, Jan. 21, 2008.

[189] E-mail entitled "RE: restructured futuregen international draft – comments requested" from Jarad Daniels to Bud Albright, James Slutz, Doug Schwartz, Adams Ingols, Kathy Fredriksen, Diana Clark and Raj Luhar, Jan. 25, 2008. At this point, India had contributed $4 million and South Korea had contributed $2 million. China and Australia had made formal commitments; Norway was ready to contribute funds; and Italy and Poland had stated interest. *Ibid.*

[190] "Draft Email from Jim Slutz to Staff Contacts in seven FG partner countries," undated.

[191] Letter from Secretary Bodman to the Honorable Akira Amari, Feb. 1, 2008.

I am really surprised that I had no prior explanation of that restructuring intention from DOE before . . . Korea really tried our best to cooperate with US to develop FutureGen project since early 2006 . . . . We contributed $2 million in March 2007 . . . . actively participated in four meetings . . . even hosted the third negotiating meeting for the FutureGen project agreement last October in Seoul to make the project move on. If you have recognized all Korea's endeavor regarding the project, it is not the appropriate way to deliver US DOE's intention to restructure FutureGen project by sending me an e-mail . . . without any prior consultation or explanation to Korea.

Mr. Kim concluded by pointedly noting "that there were better ways (both procedure and timing) to inform Korea US DOE's intention to restructure FutureGen project."[192]

After the announcement, the State Department asked if DOE had talking points to use with foreign audiences. Norway and Russia had expressed interest in FutureGen; other embassies had pro-FutureGen points in their standard talks on energy and climate.[193] On February 1, 2008, David Mulford, the U.S. ambassador to India, wrote Secretary Bodman expressing concern about the FutureGen project based on his reading of media reports. "Since I will have to address the issue soon with the Government of India (GOI) and the Indian media, I would appreciate some clarification . . . . This would include the specific issue of the status of India's pledged monetary commitment." The ambassador reiterated India's ambitious plans to expand its all coal-fired thermal capacity and asked the Secretary for his views "on how to continue cooperation with India in clean-coal power generation technology and mitigation of related carbon emissions."[194] Australia also wondered what was up. "The restructuring of FutureGen has been a hot topic for our media," Australia's clean coal manager in the Department of Resources Energy and Tourism wrote. We have also been fielding representations from our own industry including companies involved in the FutureGen Alliance . . . . [W]e need to get a better understanding of what this means in terms of the International Partnership and the associated agreement being negotiated with other Governments."[195]

In February, Secretary Bodman received a letter from the Australian minister for resources, energy and tourism, who – based on the September 4, 2007, joint statement by Prime Minister Howard and President Bush – was looking forward to "a program of consultation at both the government and industry level including the means by which information on technological advances will be shared."[196] Secretary Bodman responded with a letter stating that DOE "will continue to keep you informed of significant developments in the FutureGen program and look forward to future collaborations with Australia."[197] That appears to have been the end of any real effort for international cooperation on FutureGen, once a "core objective" of the project, although FE attempted through the spring to gin up interest. Its staff made presentations to various embassies claiming that the international component was a "key priority" in the

---

[192] E-mail entitled "Re: DOE Announces Restructured FutureGen" from Kijune Kim to James Slutz, Feb. 4, 2008.
[193] E-mail entitled "FutureGen Talking Points" from Peter Haymond to Giulia Bisconti, Jan. 31, 2008.
[194] Letter from Mr. Mulford to Secretary Bodman, Feb. 1, 2008.
[195] E-mail entitled "Re: FutureGen [SEC=UNCLASSIFIED]" from John Karas to Victor Der, Feb. 8, 2008.
[196] Letter from Martin Ferguson to Secretary Bodman, Feb. 22, 2008.
[197] Letter from Secretary Bodman to Mr. Ferguson, March 26, 2008.

restructured FutureGen with a focus on a "non-proprietary information exchange."[198] Their objective was to convey "the clear message that the U.S. commitment to clean coal remains stronger than ever under the restructured FutureGen."[199]

By the end of June, 2008 DOE claimed that it was still "exploring ways to engage governments in deploying Near-Zero Emission Coal plants with CCS for deployment around the world." It proposed workshops and symposia to share non-proprietary information and the development of global outreach strategies for acceptance of the technology and gamely claimed that all of the previously interested countries would "likely have continued interest" in the outcome of FutureGen.[200] Jim Connaughton, CEQ chief and loyal Bush soldier, was quoted in the Indian press as saying that there would be 3-4 zero emission coal-fired power plants and even greater international participation in the restructured FutureGen, although there was no evidence that either one of those statements was accurate.[201]

Australia, however, went ahead on its own. After the fall of the Howard government, it ratified the Kyoto Protocol and established its own fund to pursue CCS demonstration projects in Australia.[202]

Peabody Energy, one of the FutureGen partners which already had a presence in China, signed an agreement in December of 2007 with China Huaneng Group to invest in an integrated gasification combined cycle power plant near Tianjin, southeast of Beijing called GreenGen, although there will be no CCS until its "later phases."

Abu Dhabi is designing an IGCC plant with BP and Rio Tinto that is supposed to produce hydrogen for energy and $CO2$ to be sequestered.[203]

Conclusion

FutureGen began life as the centerpiece of the Bush Administration's climate change technologies. This initiative held out the promise of reducing greenhouse gas emissions without the pain of signing up to the Kyoto Protocols. In abandoning the original concept, the Department of Energy left the country with no coherent strategy for carbon capture and

[198] "FutureGen – International Component," attached to e-mail entitled "FW: FutureGen: International" from Victor Der to Jarad Daniels, Joseph Giove and Samuel Biondo, May 20, 2008.

[199] *Ibid.*

[200] DOE Office of Fossil Energy, "U.S. Carbon Capture and Storage Program: Where We Are and Where We're Going: Clean Coal, FutureGen, and CCS" and attachments, June 2008. This presentation was created by FE as part of a FutureGen "outreach and communications" strategy after a *New York Times* article said the entire clean coal effort was stalled. "Mounting Costs Slow the Push for Clean Coal," *The New York Times,* May 30, 2008, A1. "We will tout our investment and accomplishments as Connaughton has delineated and work them into the FE Clean Coal Exhibit," FE staff wrote. They would also visit the science attaches at the embassies in Washington and tell them about the restructured FutureGen. E-mail from Samuel Biondo to Joseph Giove, May 30, 2008.

[201] "Commercial viability of FutureGen to be known only in 2020," *The Hindu Business Line,* http://www.thehindubusinessline.com/2008/06/18/stories/2008061851582100.htm June 18, 2008.

[202] "Remember FutureGen?" *Columbia Journallism Review,* April 4, 2008; "Investment in Victoria's Clean Coal Industry," http://www.investvictoria.com/300408VicCleanCoalIndustry, April 30, 2008.

[203] "BP Says Abu Dhabi Hydrogen-Fueled Plant to Start 2013," Bloomberg.com, Jan 19, 2009. http://www.bloomberg.com/apps/news?pid=20601130&sid=azs2rxpX__Sk&refer=environment

sequestration—despite having fingers in many pots. Whether the new Administration and Congress should revive the original program, which was ready to begin work when the Department of Energy killed it, or move to some other initiative, is an open question. It is absolutely clear that the "Plan B" initiative sold to the public and the Congress by Secretary Bodman will not provide the kind of long-term benefits to the United States and the world needed to deal with global climate change. The end result of this trail of mismanagement? Progress on the great challenges to harness technology to build a greener energy future was stalled, and the United States abandoned its global leadership role.

This is a disappointing legacy for the Department of Energy.

www.ingramcontent.com/pod-product-compliance
Lightning Source LLC
Chambersburg PA
CBHW082032190526
45166CB00017B/3196